Hydrology, Ecology, and Fishes of the
Klamath River Basin

Committee on Hydrology, Ecology, and Fishes of the Klamath River

Board on Environmental Studies and Toxicology

Water Science and Technology Board

Division on Earth and Life Studies

NATIONAL RESEARCH COUNCIL
OF THE NATIONAL ACADEMIES

THE NATIONAL ACADEMIES PRESS
Washington, D.C.
www.nap.edu

THE NATIONAL ACADEMIES PRESS   500 Fifth Street, NW   Washington, DC 20001

NOTICE: The project that is the subject of this report was approved by the Governing Board of the National Research Council, whose members are drawn from the councils of the National Academy of Sciences, the National Academy of Engineering, and the Institute of Medicine. The members of the committee responsible for the report were chosen for their special competences and with regard for appropriate balance.

This project was supported by Contract No. 05CS811145 between the National Academy of Sciences and the U.S. Bureau of Reclamation. Any opinions, findings, conclusions, or recommendations expressed in this publication are those of the author(s) and do not necessarily reflect the view of the organizations or agencies that provided support for this project.

International Standard Book Number-13: 978-0-309-11506-3 (Book)
International Standard Book Number-10: 0-309-11506-X (Book)
International Standard Book Number-13: 978-0-309-11507-0 (PDF)
International Standard Book Number-10: 0-309-11507-8 (PDF)
Library of Congress Control Number: 2008921799

Additional copies of this report are available from

The National Academies Press
500 Fifth Street, NW
Box 285
Washington, DC 20055

800-624-6242
202-334-3313 (in the Washington metropolitan area)
http://www.nap.edu

Cover: Design by Liza Hamilton, National Research Council. Image of the Klamath River, courtesy of Stephan McMillan (*http://www.sonic.net/aquatint/index.html*).

THE NATIONAL ACADEMIES
Advisers to the Nation on Science, Engineering, and Medicine

The **National Academy of Sciences** is a private, nonprofit, self-perpetuating society of distinguished scholars engaged in scientific and engineering research, dedicated to the furtherance of science and technology and to their use for the general welfare. Upon the authority of the charter granted to it by the Congress in 1863, the Academy has a mandate that requires it to advise the federal government on scientific and technical matters. Dr. Ralph J. Cicerone is president of the National Academy of Sciences.

The **National Academy of Engineering** was established in 1964, under the charter of the National Academy of Sciences, as a parallel organization of outstanding engineers. It is autonomous in its administration and in the selection of its members, sharing with the National Academy of Sciences the responsibility for advising the federal government. The National Academy of Engineering also sponsors engineering programs aimed at meeting national needs, encourages education and research, and recognizes the superior achievements of engineers. Dr. Charles M. Vest is president of the National Academy of Engineering.

The **Institute of Medicine** was established in 1970 by the National Academy of Sciences to secure the services of eminent members of appropriate professions in the examination of policy matters pertaining to the health of the public. The Institute acts under the responsibility given to the National Academy of Sciences by its congressional charter to be an adviser to the federal government and, upon its own initiative, to identify issues of medical care, research, and education. Dr. Harvey V. Fineberg is president of the Institute of Medicine.

The **National Research Council** was organized by the National Academy of Sciences in 1916 to associate the broad community of science and technology with the Academy's purposes of furthering knowledge and advising the federal government. Functioning in accordance with general policies determined by the Academy, the Council has become the principal operating agency of both the National Academy of Sciences and the National Academy of Engineering in providing services to the government, the public, and the scientific and engineering communities. The Council is administered jointly by both Academies and the Institute of Medicine. Dr. Ralph J. Cicerone and Dr. Charles M. Vest are chair and vice chair, respectively, of the National Research Council.

www.national-academies.org

COMMITTEE ON HYDROLOGY, ECOLOGY, AND FISHES OF THE KLAMATH RIVER

Members

WILLIAM GRAF (Chair), University of South Carolina, Columbia
MICHAEL CAMPANA, Oregon State University, Corvallis
GEORGE MATHIAS KONDOLF, University of California, Berkeley
JAY LUND, University of California, Davis
JUDITH MEYER, University of Georgia, Athens
DENNIS MURPHY, University of Nevada, Reno
CHRISTOPHER MYRICK, Colorado State University, Fort Collins
TAMMY NEWCOMB, Michigan Department of Natural Resources, Lansing
JAYANTHA OBEYSEKERA, South Florida Water Management District, West
    Palm Beach
JOHN PITLICK, University of Colorado, Boulder
CLAIR STALNAKER, U.S. Geological Survey (retired), Fort Collins, CO
GREGORY WILKERSON, University of Illinois, Urbana
PRZEMYSLAW (ANDY) ZIELINSKI, Ontario Power Generation, Toronto

Staff

DAVID POLICANSKY, Study Director
SUZANNE VAN DRUNICK, Senior Program Officer
LAUREN ALEXANDER, Senior Program Officer
RUTH CROSSGROVE, Senior Editor
MIRSADA KARALIC-LONCAREVIC, Manager of the Technical Information
    Center
JORDAN CRAGO, Senior Project Assistant
RADIAH ROSE, Senior Editorial Assistant

Sponsor

U.S. BUREAU OF RECLAMATION

SUSAN N.J. MARTEL, Senior Program Officer for Toxicology
KULBIR BAKSHI, Senior Program Officer
KARL E. GUSTAVSON, Senior Program Officer
ELLEN K. MANTUS, Senior Program Officer
RUTH E. CROSSGROVE, Senior Editor

Strengthening Science at the U.S. Environmental Protection Agency (2000)

Scientific Frontiers in Developmental Toxicology and Risk Assessment (2000)

Ecological Indicators for the Nation (2000)

Waste Incineration and Public Health (2000)

Hormonally Active Agents in the Environment (1999)

Research Priorities for Airborne Particulate Matter (four volumes, 1998-2004)

The National Research Council's Committee on Toxicology: The First 50 Years (1997)

Carcinogens and Anticarcinogens in the Human Diet (1996)

Upstream: Salmon and Society in the Pacific Northwest (1996)

Science and the Endangered Species Act (1995)

Wetlands: Characteristics and Boundaries (1995)

Biologic Markers (five volumes, 1989-1995)

Review of EPA's Environmental Monitoring and Assessment Program (three volumes, 1994-1995)

Science and Judgment in Risk Assessment (1994)

Pesticides in the Diets of Infants and Children (1993)

Dolphins and the Tuna Industry (1992)

Science and the National Parks (1992)

Human Exposure Assessment for Airborne Pollutants (1991)

Rethinking the Ozone Problem in Urban and Regional Air Pollution (1991)

Decline of the Sea Turtles (1990)

Copies of these reports may be ordered from
the National Academies Press
(800) 624-6242 or (202) 334-3313
www.nap.edu

OTHER REPORTS OF THE WATER SCIENCE
AND TECHNOLOGY BOARD

Colorado River Basin Water Management: Evaluating and Adjusting to Hydroclimatic Variability (2007)

Improving the Nation's Water Security: Opportunities for Research (2007)

CLEANER and NSF's Environmental Observatories (2006)

Drinking Water Distribution Systems: Assessing and Reducing Risks (2006)

Progress Toward Restoring the Everglades: The First Biennial Review, 2006 (2006)

River Science at the U.S. Geological Survey (2006)

Second Report of the National Academy of Engineering/National Research Council Committee on New Orleans Regional Hurricane Protection Projects (2006)

Structural Performance of the New Orleans Hurricane Protection System During Hurricane Katrina: Letter Report (2006)

Third Report of the NAE/NRC Committee on New Orleans Regional Hurricane Protection Projects (2006)

Toward a New Advanced Hydrologic Prediction Service (AHPS) (2006)

Public Water Supply Distribution Systems: Assessing and Reducing Risks (2005)

Re-engineering Water Storage in the Everglades: Risks and Opportunities (2005)

Regional Cooperation for Water Quality Improvement in Southwestern Pennsylvania (2005)

Review of the Lake Ontario-St. Lawrence River Studies (2005)

Science of Instream Flows: A Review of the Texas Instream Flow Program (2005)

Water Conservation, Reuse, and Recycling (2005)

Water Resources Planning for the Upper Mississippi River and Illinois Waterway (2005)

Adaptive Management for Water Resources Project Planning (2004)

Analytical Methods and Approaches for Water Resources Project Planning (2004)

Army Corps of Engineers Water Resources Planning: A New Opportunity for Service (2004)

Assessing the National Streamflow Information Program (2004)

Confronting the Nation's Water Problems: The Role of Research (2004)

Contaminants in the Subsurface: Source Zone Assessment and Remediation (2004)

Groundwater Fluxes Across Interfaces (2004)

Managing the Columbia River: Instream Flows, Water Withdrawals, and Salmon Survival (2004)

Copies of these reports may be ordered from
the National Academies Press
(800) 624-6242 or (202) 334-3313
www.nap.edu

Preface

The Klamath River basin is both at the edge and at the center. The basin is a 15,700 square mile watershed at the western rim of North America, where it encompasses a diverse ecosystem, wilderness areas, and irrigated farmlands in southern Oregon and Northern California. The basin is located at the center, however, of the landscape of controversy in American environmental management, and the issues that face Klamath River basin decision makers exemplify in magnified form many of the difficult science and policy challenges that have arisen across the continent. Management of the basin's hydrologic and ecological resources is complicated because decision makers must sort through a myriad of potential strategies for operating a complex system with interrelated rivers, lakes, marshes, dams, and diversions. The river basin boundaries outline an ecosystem that includes economically valuable water resources and ecologically valuable species, including endangered, threatened, and other fishes, which are dependent on the rivers and lakes for their survival. Alterations to the original hydrologic system began in the late 1800s, accelerated in the early 1900s, and continue today. They include water-control works by private land and water owners, by the large and intricate Klamath Irrigation Project of the U.S. Bureau of Reclamation (USBR), and by several hydroelectric dams operated by a private corporation, PacifiCorp.

These hydrologic alterations, combined with overfishing, habitat alteration, poor water quality, and nonnative species, have led to a dramatic decline in coho salmon, Lost River suckers, and short-nose suckers and some other fishes of the Klamath River. Salmon, once providing the basis of the third largest salmon fishery among west-coast rivers, are a critical component of the ecosystems and cultural systems of the Klamath region. By the end of the twentieth century, the inherent difficulties in balancing the benefits of the river's water for fish, agriculture, and hydropower had become further complicated by national resource policies supporting Native American rights, water development, hydropower production, and endangered and threatened species.

Science and engineering have been the handmaidens of water development in the Klamath River basin, and decision makers have called upon science and engineering expertise to aid them in sorting out the choices for future management of the basin's water and water-related resources. Recognizing that the best decisions are likely to benefit from the understanding derived from scientific research and engineering investigations, in 2001 the U.S. Department of Interior and the U.S. Department of Commerce requested that the National Research Council (NRC) form a committee to complete two reports. The first (interim) study, completed in 2002, assessed the strength of scientific support for the 2001 biological assessments and biological opinions on the three endangered or threatened fish species in the Klamath River basin. The second (final) study, completed in 2004, evaluated the 2002 biological assessments and biological opinions, and other matters related to the long-term survival and recovery of the federally listed fish species.

Subsequently, in 2005 the Bureau of Indian Affairs (on behalf of the Native American tribes of the basin) and the USBR (serving many irrigators in the basin) requested that the NRC conduct a more specific evaluation and review two new studies, completed after 2004, which were designed to inform decision makers about the hydrology and fish ecology of the Klamath River basin. In order to define hydrologic conditions that supported the predevelopment fish population, one study used data and modeling approaches to gain an understanding of what the natural flows of the river might be without the presence of agriculture and the water control infrastructure. The second study created a model-based linkage between the hydrology and the resulting aquatic ecosystems that support the fish populations in the river. The present report is the outcome of the NRC review and evaluation of those studies.

The committee is grateful for the support of USBR officials Pablo Arroyave, William Rinne, James Hess, and William Shipp. Many people with close associations with the Klamath River basin aided the committee in its efforts to understand the Klamath River basin and its resources. The people of Yreka, California, and Klamath Falls, Oregon, made the committee welcome and shared their perspectives during committee visits to those communities. During public sessions associated with those visits, local citizens joined federal, state, and local agency representatives in discussions and presentations for the committee. Jon Hicks and Cindy Williams, of the

USBR's Klamath Falls Office, were particularly helpful to the committee in gaining an understanding of the Klamath Project, a key component of the present basin system. During a visit by some committee and staff members to Utah State University in Logan, Dr. Thomas Hardy extended every courtesy, as did Craig Albertson, Elizabeth Cohen, Alan Harrison, Thomas Perry, and Mark Spears during another similar visit to the USBR offices in Denver, Colorado. These researchers repeatedly aided the committee in tracking down information, data, and elusive documents.

The committee also benefited from terrific support from the NRC staff. James Reisa (director of the Board on Environmental Studies and Toxicology) and Stephen Parker (director of the Water Science and Technology Board) provided a supportive institutional home for the committee and its members. David Policansky (scholar and senior program officer of the Board on Environmental Studies and Toxicology) played a pivotal role in the deliberations of the committee and the writing of the report. His wide experience, range of knowledge, and congenial interactions with the committee were important contributions to the result. Suzanne van Drunick (project director and senior program officer) guided the committee with great wisdom and adroit management through its meetings and its report writing, providing organizational skills and knowledge of the Klamath issues that made the report possible. The extensive hydrologic knowledge and sound judgment of Lauren Alexander (senior staff officer of the Water Science and Technology Board) contributed substantially to early stages of development of the committee and its report. The complicated mechanics of arrangements for committee meetings and travel, as well as the smooth production of the meetings was in the capable hands of senior program assistants Liza Hamilton and Jordan Crago. Ruth Crossgrove did her usual scholarly job of editing the report. Thank you to all of these talented NRC professionals.

This report is the consensus expression of the committee's conclusions and recommendations, but it is actually the product of hard work and thoughtful review. We express our appreciation to members of the Board on Environmental Studies and Toxicology and the Water Science Board; to the NRC's Report Review Committee, which took on the responsibility of external review oversight; and to the independent scientists and engineers listed below, who reviewed the report. These reviewers provided us with insightful commentary, numerous penetrating questions, and exceptionally helpful suggestions for clarifying and improving our report. We benefited enormously from their help.

This report has been reviewed in draft form by individuals chosen for their diverse perspectives and technical expertise, in accordance with procedures approved by the NRC's Report Review Committee. The purpose of this independent review is to provide candid and critical comments that will assist the institution in making its published report as sound as possible and to ensure that the report meets institutional standards of objectivity, evidence, and responsiveness to the study charge. The review comments and draft manuscript remain confidential to protect the integrity of the deliberative process. We thank the following individuals for their review of this report:

Stanley Gregory, Oregon State University
Robert Huggett, Seaford, Virginia
William Lewis, University of Colorado
David Maidment, University of Texas
Jeffrey Mount, University of California at Davis
Patrick O'Brien, ChevronTexaco Energy Technology Company
LeRoy Poff, Colorado State University
Gordon Robilliard, Entrix, Inc.
Kenneth Rykbost, Klamath Falls, Oregon

Although the reviewers listed above have provided many constructive comments and suggestions, they were not asked to endorse the conclusions or recommendations, nor did they see the final draft of the report before its release. The review of this report was overseen by the review coordinator, Paul G. Risser, of Oklahoma State University, and the review monitor, Gordon H. Orians, of the University of Washington (emeritus). Appointed by the NRC, they were responsible for making certain that an independent examination of this report was carried out in accordance with institutional procedures and that all review comments were carefully considered. Responsibility for the final content of this report rests entirely with the authoring committee and the institution.

To my fellow committee members, I express a special debt of gratitude. They were a committee drawn from different backgrounds and disciplinary cultures, yet they were willing to work together in a harmonious collective effort to address the complexities of science and engineering for the Klamath River system. They put aside their personal biases, worked long hours that sacrificed their own professional time, and traveled great distances to make their contributions to this report. Such unpaid service is remarkable, but the committee received a truly remarkable recompense: the opportunity to contribute the experience and knowledge collected from our careers to support a public vision for the future of the basin and its resources. It is our hope that, although the Klamath River basin is at the edge of the continent, it will also be a central example of successful application of science and engineering to American ecosystem restoration and management.

William L. Graf, *Chair*
Committee on Hydrology, Ecology, and
Fishes of the Klamath River Basin

Contents

BOXES, TABLES, AND FIGURES

BOXES

TABLES

FIGURES

Hydrology, Ecology, and Fishes of the
Klamath River Basin

Summary

The Klamath basin of northern California and southern Oregon has been the scene of controversies over water allocations in recent years. As often is the case with environmental controversies, a considerable amount of science has been done in the basin. However, the continuing lack of an overall model or vision to provide a framework for identifying science needs has prevented the science from being used effectively enough in decision making and management to resolve the continuing controversies, which have led to the involvement of the National Research Council (NRC). This report, which has as its main focus review of two large efforts to model the hydrology of the basin (the Natural Flow Study by the U.S. Bureau of Reclamation) and the relationship of Klamath River hydrology to habitat for salmon (by Utah State University), also addresses the broader questions of the ecological needs of the anadromous fishes and the importance of a broad, comprehensive view of the basin's science needs as a guide to scientific activities.

The Klamath basin has been extensively modified by levees, dikes, dams, and the draining of natural water bodies since the Klamath Project was begun in 1905 to improve the region's ability to support agriculture; other changes have occurred as well. All those changes have been accompanied by changes in the biota of the basin. Of particular concern in this report are changes in the distribution and abundance of several species of fishes in the Klamath River and in its tributaries. Those fishes were the subject of earlier NRC reviews prompted by conflicts that arose after management actions were taken to protect the basin's fishes during the very dry year of 2001; one result of those actions was a severe reduction in the water available for agriculture. In addition, in September of 2002, more than 33,000[1] mostly adult fish died in the lower Klamath River, about 95% of which were Chinook salmon, the remainder being mostly steelhead. Less than 1% of the deaths were coho salmon. This mass mortality intensified the controversy over water operations in the Klamath basin.

The management and uses of the natural resources of the basin, including water and fishes, are complex. Many federal, state, county, and other agencies and organizations are involved, and the basin's resources are managed to achieve a variety of divergent purposes.

RECENT EVENTS LEADING TO THIS STUDY

The Endangered Species Act requires that the U.S. Bureau of Reclamation (USBR) make assessments of the effects of the Klamath Project operations on fishes listed as threatened or endangered and consult about those assessments with the U.S. Fish and Wildlife Service (USFWS) for Lost River and shortnose suckers in Klamath Lake and the National Marine Fisheries Service (NMFS) for coho salmon in the lower Klamath River. The assessments that led to the NRC study initially were conducted in 2001. After consultations, the USFWS endorsed some of the USBR proposals, but concluded that more water than the USBR proposed was needed to maintain Upper Klamath Lake at levels that would protect the suckers. The NMFS also agreed with some of the USBR proposals, but concluded that more water was needed to maintain higher minimum flows in the Klamath River below Iron Gate Dam than the USBR had proposed. The "biological opinions" of the USFWS and the NMFS indicated that some of the USBR's proposals would jeopardize the continued existence of the listed species, and therefore the USBR was required to allot more water to the lake and to the river than had been planned, leaving less water than had previously been allocated for agriculture.

Those restrictive allocations, coupled with a very dry year, resulted in hardship for many of the basin's water users, and the controversy surrounding the allocations became intense. As a result of the controversy, the U.S. Department of the Interior asked the NRC to review the scientific bases of the USBR biological assessments and the USFWS and NMFS biological opinions. In response, the NRC established the Committee on Endangered and Threatened Fishes in the Klamath River Basin, which issued an interim report in 2002 that focused on the biological assessments and biological opinions and a final report in 2004 that took a broader look at strategies for recovery of the endangered and threatened fishes of the basin. The in-

---

[1]The California Department of Fish and Game, which made this estimate, described it as "conservative."

terim report concluded that most of the recommendations of the biological opinions had scientific support but that available scientific data did not support the higher minimum lake levels or the higher minimum river flows recommended in the biological opinions to benefit the species listed under the Endangered Species Act. The later report confirmed those conclusions and included many recommendations for actions to benefit the listed fish species and to improve scientific understanding of the basin.

Since the publication of the NRC reports, two new documents have become available: an estimate of natural or unimpaired flows in the basin as they were before the project was begun (the Natural Flow Study) and a model of the relationship of flows in the Klamath River to habitat in the river available for endangered and threatened fishes there, especially coho salmon (often referred to as Hardy Phase II, referred to here as the Instream Flow Study Phase II). Because those new documents have the potential to change scientific conclusions and management options based on earlier information, the Department of the Interior asked the NRC to evaluate them and their implications for the biota of the basin. In response, the NRC established the Committee on Hydrology, Ecology, and Fishes of the Klamath River Basin, which prepared this report. New developments have occurred since the previous reports were published, so this report is not a revisiting of the issues covered by the earlier ones. This committee endorses the recommendations of the earlier reports for reversing the declines of the listed species, and this report should be considered as building on the previous ones, continuing where they left off.

THE PRESENT STUDY

Statement of Task

A multidisciplinary committee will be established to evaluate new scientific information that has become available since the NRC issued its 2004 report on Endangered and Threatened Fishes in the Klamath River Basin. The new information to be evaluated by the committee will include two new reports on (1) the hydrology of the Klamath Basin and (2) habitat needs for anadromous fish in the Klamath River, including coho salmon. The committee will also identify additional information needed to better understand the basin ecosystem.

To complete its charge, the committee will

1. Review and evaluate the methods and approach used in the Natural Flow Study to create a representative estimate of historical flows and the Hardy Phase II studies, to predict flow needs for coho and other anadromous fishes.

2.   Review and evaluate the implications of those studies' conclusions within the historical and current hydrology of the upper basin; for the biology of the listed species; and separately for other anadromous fishes.

3.   Identify gaps in the knowledge and in the available scientific information.

Committee Process

To execute its charge, the committee met four times. At the first three meetings, the committee heard presentations from scientists and others, including agency officials familiar with various aspects of the region and the operation of the Klamath Project; the committee also received presentations from the public. The committee visited a restoration and research project on the upper Shasta River, the Iron Gate Dam and hatchery on the Klamath River, and the monitoring station near the mouth of the Shasta River. Individual members of the committee and staff also visited other parts of the basin, including portions of the Klamath River downstream of Iron Gate Dam, Upper Klamath Lake; the Williamson, Sprague, and Wood Rivers; the Link River and Link River Dam; Keno Dam, and J.C. Boyle Dam.

CONCLUSIONS AND RECOMMENDATIONS

We present the committee conclusions on the Natural Flow Study and the Instream Flow Study Phase II, along with recommendations for their improvement, followed by more general conclusions and recommendations for the conduct of science for management in the Klamath basin. The committee concludes that a more coherent, systematic, and comprehensive analysis of scientific and management needs for the basin should be conducted to identify the most important and urgent science needs to inform management decisions. Only when—and if—that analysis concludes that the Natural Flow Study and the Instream Flow Study Phase II are important components of such a comprehensive framework should the committee's recommended improvements to them be implemented.

THE NATURAL FLOW STUDY

USBR conducted the study *Natural Flow of the Upper Klamath River* to "estimate the effects of agricultural development on natural flows in the Upper Klamath River Basin" using an "estimate of the monthly natural flows in the Upper Klamath River at Keno." Essentially, the USBR study provided flow estimates that would be observed if there *were no agricultural development, such as draining of marshes and diversions of flow,* in the upper Klamath basin. The products of the study were to be used as inputs for the Instream Flow Study. The study and the committee's evaluation of it are described in detail in Chapter 4.

Committee Evaluation

The Natural Flow Study for the Klamath River has several admirable attributes. The data sets describing stream flow that the Natural Flow Study assembled are extensive and are highly useful. The conceptual model developed to identify the components needed in a natural-flow model appears to be adequate. The simulated data adequately reflect the monthly seasonality of the flow system. Human activities have modified that system over substantial portions of the basin above the Iron Gate Dam gauge site, and USBR investigators included many of these modifications in their calculations. Investigators recognized the importance of marsh conversions and agricultural activities in affecting river flows, and included these factors in their calculations. The documentation for the Natural Flow Study is accessible to the reader and provides a straightforward explanation of what the modelers did and how they did it, and provides the complete output of the research. The report also addresses important issues about the natural flow model, including brief accountings of model verification, sensitivity, and uncertainty.

The committee concluded, however, that the Natural Flow Study was seriously compromised by several fundamental issues, including its choice of a basic approach for understanding natural flows, choices of the models for calculations, and serious omissions of factors likely to influence river flows at the Iron Gate Dam gauge site, as described below:

- The products of the Natural Flow Study, flow values for the Klamath River at the Iron Gate Dam site, were calculated as monthly values. The ecological applications of these calculated flows require daily values, and as a result, the output of the study would not have satisfied its ultimate use requirements even if the study had been executed without other errors.
- The USBR researchers relied on a "black box"[2] method of accounting for flow using a standard spreadsheet as the foundation. The U.S. Geological Survey's (USGS's) Modular Modeling System (MMS) provides greater flexibility and adaptability and provides a firmer theoretical foundation than a straightforward accounting system.
- The calculations of the fate of water in the upper basin related to evapotranspiration were not done according to the best current methods,

---

[2]A "black box" method attempts to investigate a complex process—in this case, flows—without making assumptions about the mechanisms or structures that affect the process.

such as the Food and Agriculture Organization's[3] (FAO's) version of the modified Blaney-Criddle method. A more serious concern was the model behavior when a sensitivity analysis of its output concerning agricultural land was conducted. The results were not explained, and the apparent anomaly appears to be related to the component of the model that deals with reduction of evapotranspiration in the Upper Klamath Lake marsh when it is converted to agriculture.

• The USBR attempted to calculate flows at Iron Gate Dam without adequately addressing important controlling factors for those flows, including groundwater.

• More generally, the Natural Flow Study did not fully address the issue of changes in land use and land cover. The inclusions of land-use and land-cover analyses in the study would have increased confidence in the resulting calculations because, if such changes are important, they would reflect their influence in the model output; if the changes are unimportant, that outcome could be convincingly demonstrated.

• The study failed to adequately model the connection between the Klamath River and Lower Klamath Lake.

• The study did not adhere closely enough to standard scientific and engineering practice in the areas of calibration, testing, quality assurance, and quality control. For example, the natural-flow model cannot be calibrated using standard modeling practices. A reasonable check on the model can be made only by using the data from the earliest available measurements of flows.

The committee concluded that the Natural Flow Study includes calculated flows that are at best first approximations to useful estimates of such flows. The present version of the Natural Flow Study is less than adequate for input to the Instream Flow Study Phase II and does not provide enough information for detailed management of flows for the benefit of listed and other anadromous fish species in the Klamath River downstream from Iron Gate Dam. However, it does provide some basis for understanding unimpaired flows in the basin and for providing a context for more detailed management decisions. To become useful for more precise decision making in daily or even monthly flow management, the Natural Flow Study should be improved by (1) replacing the Soil Conservation Service (SCS)[4] modified Blaney-Criddle method for calculating evapotranspiration with a more accurate and modern version, such as the FAO version of the method, using generally available data; (2) including groundwater dynamics in the model

---

[3]An agency of the United Nations.
[4]An agency of the U.S. Department of Agriculture, now called the Natural Resources Conservation Service.

in at least a general way; (3) improving the portions of the predictive model relating to land use and land cover so that changes in these variables are represented in a more complete fashion; (4) including the role of the Lost River and Lower Klamath Lake in the complicated high-flow scenarios; (5) replacing the black-box accounting method based on a spreadsheet with a more robust physically based model for generating flows, such as the USGS's MMS, or its new model GSFLOW, which combines the MMS with the groundwater model MODFLOW; (6) including an extensive investigation of high flows along with their geomorphic and ecological implications; and (7) adhering more closely to standard scientific and engineering practice by extensively calibrating and testing the models and their underlying software, while addressing issues of quality assurance and quality control. The Natural Flow Study also should be modified to better meet the needs of the Instream Flow Study.

Although the Natural Flow Study has advanced our understanding of the basin, its weaknesses also point to next steps that would help development of hydrologic models better suited and more transparent for the basin's current problems.

INSTREAM FLOW STUDY PHASE II

The Instream Flow Study Phase II for the Klamath River Basin accepted information from the Natural Flow Study discussed above and produced recommendations for instream flows at the USGS stream gauge below Iron Gate Dam. To reach those recommendations, the Instream Flow Study Phase II included an elaborate series of investigations and model-building efforts. The general technical elements of an instream flow study, the procedures followed in this particular case, and the committee's evaluation of those procedures are described in detail in Chapter 5.

Committee Evaluation

Several aspects of the Instream Flow Study Phase II are praiseworthy. The measurement of stream-bed topography and of substrate characteristics in this study represents innovative cutting-edge methods that provided generally useful representations of the river channels. The two-dimensional hydrodynamic model in the Instream Flow Study Phase II represented the state-of-the-art application of flow models in simulating habitats. The application of two-dimensional approaches represented a willingness on the part of the investigators to engage in a highly complex and ambitious project to deal with the hydraulic and hydrologic aspects of the problem of characterizing fish habitat. The study incorporated distance to escape cover, an important variable that is sometimes ignored in other studies.

As a general perspective, the Instream Flow Study Phase II followed steps outlined in the Instream Flow Incremental Methodology (IFIM), which has seen wide application in studies of this type. The authors of the Instream Flow Study Phase II applied the IFIM properly. They also used bioenergetics and a fish-population model to test their results, and they tested model output by comparing observations of fish with predicted fish locations.

Despite these strengths, the committee found important shortcomings in the Instream Flow Study Phase II and its use of various models and data. Two major shortcomings—use of monthly data and lack of tributary analyses—are so severe that that they should be addressed before decision makers can use the outputs of the study to establish precise flow regimes with confidence. Neither was the fault of the authors of the Instream Flow Study Phase II; the shortcomings resulted from constraints imposed by the USBR, which indicated that lack of time and resources prevented them from providing additional calculations that would produce daily flows for the ecological modeling. Although monthly flow values can be useful for general river-basin planning, they are not useful for ecological modeling for river habitats, because the monthly average masks important discharge values that may exist only for a few days or even less. In short, planners operate on a monthly basis, but fish live on a daily basis.

The elimination of consideration of tributary processes apparently resulted from an agreement reached by basin managers not to include tributary processes in the habitat studies to simplify the engagement of stakeholders in the process. Since only the main stem of the Klamath River was subject to analysis, stakeholders with interests in tributary locations would not have to deal directly with the study. The Klamath River is not a confined gutter for rainwater, and therefore analyzing the river without considering its tributaries is akin to analyzing a tree by assessing only its trunk but not its branches. In addition, the study did not include important water-quality attributes, such as dissolved oxygen levels, nutrient loadings, contaminants, and sediment concentrations, although each has important implications for the vitality of the fish populations of the Klamath River. Second, high flows are especially important to the physical and biological processes of the Klamath River, and further analysis of their frequency, duration, and timing is essential in understanding the dynamics of the river's hydrologic, geomorphologic, and ecological processes. Reliance on monthly flow data, as outlined above, made analysis of high flows impossible within the scope of the study.

Third, there was a lack of a thorough assessment of the relationship between flow-data time series and the behavior of different species and life stages and the population dynamics of coho and Chinook salmon. Fourth, the claim that the model outcomes are accurate, as assessed by some empirical tests of fish distributions and by use of the SALMOD model, are not substantiated, impairing the utility of the Instream Flow Study Phase II. Statistical measures of the closeness of fit between model predictions and fish occurrence would substantially increase the confidence of users in the outputs of the study.

Finally, there are three major shortcomings in the experimental design of the Instream Flow Study Phase II: a fundamental beginning assumption about limits on salmon habitat, a lack of thorough assessment of the representativeness of the reaches used for detailed study, and the statistical approach to analyze the calculated set of instream flows did not use normalized data and did not have provisions for identifying serial autocorrelations.

Despite these limitations, and in the absence of any better information currently available, the committee concludes that the recommended flows resulting from the Instream Flow Study Phase II probably represent an improvement for the anadromous fishes in the Klamath River over the current flow regime. These are improvements in flow because they include intra- and inter-annual variations and probably will enhance Chinook salmon growth and young-of-the-year production. Because the study was based on three species—Chinook salmon, coho salmon, and steelhead—it is not possible to know how well the recommendations apply to any one species or to all the species as a whole. Indeed, most of the information was from Chinook salmon, which suggests that confidence in its applicability to that species would be greater than to other species. To the degree that the studies conclusions are followed, it should be on an interim basis, pending the improvements the committee outlines below and a more comprehensive and integrated assessment of the science needs of the basin as a whole.

The study would be improved for greater utility by (1) using daily flows as a basis for calculations; (2) taking into account habitats, water, and sediment contributions from tributaries; (3) specifically testing how representative the selected test reaches are of the entire river; (4) rigorous statistical testing of the model outcomes to support claims of accuracy; (5) including water-quality measures, sediment loadings, and contaminants in the modeling process; (6) including extended analyses of high-flow events; (7) exploring through thorough analysis of the habitat times series the presence or absence of any life-stage habitat limitations for a variety of species and life stages for natural and existing flows; (8) substituting another stochastic approach rather than the Periodic Autoregressive Moving Average model to analyze the statistical nature of the calculated flows; and (9) conducting sensitivity analyses using dynamic fish-population growth and production models to investigate the influence of alternative flow regimes on life cycles and stages of salmon to understand the nature of bottlenecks that can potentially constrain population growth, as well as the potential for flow-related improvements. Additional suggestions for improving the model are in Chapter 7.

IMPLICATIONS FOR ANADROMOUS
FISHES IN THE KLAMATH RIVER

The Natural Flow Study

The implications of the model investigations are mixed. From a positive perspective, the results define monthly "natural" variation that managers might reasonably expect, absent their own activities. The monthly variation depicted by the model represents a simulated picture of the conditions under which the biological community of the river evolved and provides a backdrop for assessing the degree to which the present regulated flow regime departs. The flows also provide a general view of the total amounts of water involved in the river and lake regime, with about 1.4 million acre feet annually flowing out of the lake on the average.

The Natural Flow Study reasonably captures the decadal variations in flows in the system that are likely to have occurred in the absence of upper-basin development and the installation of dams. These variations imply that some decadal fluctuation in flows is reasonable in the regulated system and that a completely unchanging regime imposed by engineering structures would not reflect the natural regime.

However, the internal workings of the model in the Natural Flow Study include several computational shortcomings that limit its use. These issues imply that the natural flow model produces results that probably cannot be used as a precise replication of natural flows and that the individual numbers generated by the study are not firm, irrefutable values. The study's shortcomings imply that managers of the biological resources of the basin may use the results of the model in a general way as a form of guidance for the broad characteristics of the natural flow regime, but they cannot use the exact values produced by the study as a template for developing a flow regime with much confidence. The model is a general representation, and because its output is in monthly time steps, it is not capable of generating the daily time step needed for a completely effective instream flow model to be used in any ecological model downstream. As described in considerations of the Instream Flow Study in Chapter 5, this limitation has a ripple effect that limits the utility of the instream flow recommendations.

Finally, the current model is severely restricted for two general reasons. First, the basin and its biota have changed so much in the past century that the implications for the fishes of restoring "natural flows" are not clear.

Second, the model does not treat the tributaries of the Klamath River, although they are and have been an essential part of the environments of the anadromous fishes. Without understanding the ecological and hydrological condition and dynamics in the tributaries, it is not possible to understand the ecological and hydrological condition and dynamics of the river.

A modified version of the Natural Flow Study model, using suggestions made in this report, could have management utility. It could be used as a template for a model of the present-day system. Such a model could be used to simulate "What if?" scenarios, test certain hypotheses, and demonstrate to stakeholders the implications of assorted management decisions and stakeholder choices. Since the Natural Flow Study model is built upon a familiar, user-friendly platform (Excel), a modified model might find wide use among stakeholders.

The Instream Flow Study

The basic conclusions of the Instream Flow Study are recommended flows expressed as monthly target values for discharges below Iron Gate Dam on the Klamath River. The most important outcome of the Instream Flow Study was that it indicated that increases in existing flows downstream from Iron Gate Dam probably would benefit fish populations through improved physical habitat associated with more water and through reduced water temperatures. If these conclusions were borne out by studies incorporating experimental flows and monitored responses, managers would be able to have greater confidence that decisions to increase flows would have a beneficial effect on anadromous fishes in the lower river. The authors of the Instream Flow Study mention two caveats, and this committee agrees with them. First, the flow recommendations apply to the needs of the anadromous fishes in the lower Klamath River, and they do not account for competing water demands for other purposes, such as agricultural needs or the needs of federally listed fishes in the upper basin. Second, the flow recommendations address the needs of all the anadromous species in the lower Klamath River. They are not targeted for any individual species (listed or otherwise), and it is not possible to evaluate the conclusions separately for individual species.

Despite various concerns about the study, it is extremely unlikely, in the committee's judgment, that following the prescribed flows of the Instream Flow Study Phase II would have adverse effects on any of the anadromous fish species. Based on general principles and the information developed in that study, following its prescribed flows probably would have some beneficial effects on the suite of anadromous fishes in the Klamath River considered as a whole, although not necessarily for every species.

DEVELOPING A COMPREHENSIVE SCIENTIFIC FRAMEWORK TO CONNECT SCIENCE AND DECISION MAKING

The committee found that science in the basin was being done by bits and pieces, sometimes addressing important questions, but not linked to other important questions and their studies. The Natural Flow Study and the Instream Flow Study Phase II were major science and engineering investigations, but the linkage of one to the other was only partially achieved. Other studies in the basin, such as the USGS's hydrologic studies in the Sprague River basin, or the extensive research in the Trinity River basin (which is part of the Klamath River basin), seem not to have had any influence on each other or on the flow studies examined in this report. The committee found that the most important characteristics of research for complex river-basin management were missing for the Klamath River: the need for a "big picture" perspective based on a conceptual model encompassing the entire basin and its many components. As a result, the integration of individual studies into a coherent whole has not taken place, and it is unlikely to take place under the present scientific and political arrangements.

To address science and management in the basin, the committee first recommends that the agencies, researchers, decision makers, and stakeholders together define basin-wide science needs and priorities. One method of achieving success in this effort would be through the establishment of an independent entity to develop an integrated vision of science needs. The body that defines this vision must be viewed by all parties as truly independent for it to be effective, unlike the Conservation Improvement Program, which, despite good intentions, appears to many people in the region as a creature of the USBR, and therefore associated with the bureau's official mandates and responsibilities. If the proposed task force reports to the secretary of the interior, rather than to any specific agency, it is more likely to avoid the appearance of being controlled by any particular agency or interest group in the basin and thus more likely to be and to appear independent. Leadership of the task force by a senior scientist who reports to the secretary would be a major step toward removing perceived biases in science and its application.

The committee concludes that when the science needs for the basin are better characterized, the individual studies necessary to create a sound science-based body of knowledge for decision makers and managers will be more easily identified. Only if this general vision and process determines that the Natural Flow Study and the Instream Flow Study Phase II might help satisfy science needs in the basin should investigators seek to address the shortcomings that the committee has identified. The Trinity River basin experience, despite some difficulties, provides a good example to follow in many aspects of the overall basin-wide effort.

Connecting effective science with successful decision making for delivering water to users, sustaining downstream fisheries, and protecting the populations of protected species has been problematic in the Klamath River basin. The Natural Flow Study and the Instream Flow Study Phase II are not likely to contribute effectively to sound decision making until political and scientific arrangements in the Klamath River basin that permit more cooperative and functional decision making can be developed. The employment of sound science will require the following elements:

• A formal science plan for the Klamath River basin that defines research activities and the interconnections among them, along with how they relate to management and policy.
• An independent mechanism for science review and management that is isolated from direct political and economic influence and that includes a lead scientist or senior scientist position occupied by an authoritative voice for research.
• A whole-basin viewpoint that includes both the upper and lower Klamath River basins with their tributary streams.
• A data and analysis process that is transparent and that provides all parties with complete and equal access to information, perhaps through an independent science advisory group.
• An adaptive-management approach whereby decisions are played out in water management with monitoring and constant assessment and with periodic informed adjustments in management strategies.

The committee recommends that the researchers, decision makers, and stakeholders in the Klamath River basin emulate their colleagues in the Trinity River basin in connecting science and decision making and that the two units coordinate their research and management for the greater good of the entire river basin.

1

Introduction

The Klamath River basin of northern California and southern Oregon has been the scene of controversies over water allocations in recent years. The basin has been extensively modified by levees, dikes, dams, diversion of tributary waters, and the draining of natural water bodies since the Klamath Project was begun in 1905 to improve the region's ability to support agriculture; other changes have occurred as well (NRC 2004a).

The changes made to the system have been accompanied by changes in the biota of the basin as well. This report particularly focuses changes in the distribution and abundance of several species of fishes of concern in the Klamath River, Upper Klamath Lake, and their tributaries. Those fishes were the subject of earlier reviews by the National Research Council (NRC 2002, 2004a), the first of which focused on specific documents related to water management in the basin and its effect on the fishes and the second of which focused on broad aspects of the basin's management and options for arresting and perhaps reversing the declines of the basin's fishes. The evaluations were prompted by conflicts that arose following actions taken to protect the basin's fishes during the very dry year of 2001. One result of those actions was a severe reduction in the water available for agriculture.

THE KLAMATH RIVER BASIN

The Klamath River basin in southern Oregon and northern California covers 40,632 km^2 (15,688 mi^2) or slightly more than 4 million hectares (10 million acres). In Oregon the basin occupies portions of Jackson, Lake, and Klamath counties; in California, it includes parts of Siskiyou, Modoc,

Trinity, Humboldt, and Del Norte counties (NRCS 2006) (Figure 1-1). Annual precipitation in the upper basin, that is, above Iron Gate Dam, averages from 68 to 69 cm (about 27 in.) (Risley and Laenen 1999) but is only about 30.5 cm (12 in.) at Klamath Falls (from Weather Underground 2007). In the lower basin, annual precipitation can exceed 255 cm (100

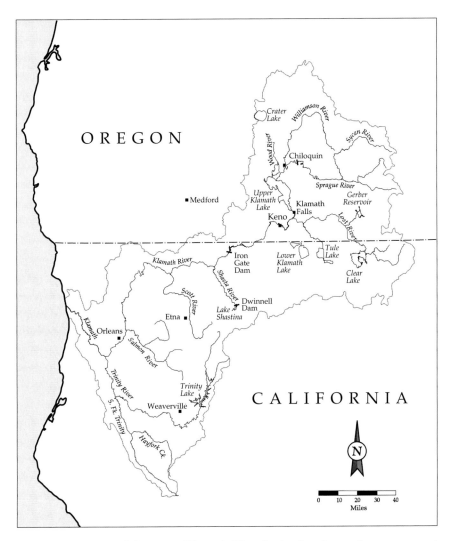

FIGURE 1-1 Map of the upper Klamath River basin showing surface waters and landmarks mentioned in this report.
SOURCES: Reproduced from NRC 2004a, modified from U.S. Fish and Wildlife Service.

in.). Above about 1,600 m (5,000 feet), large snowpacks accumulate in wet years, and runoff is high. Land elevations exceed 2,000 m to the west, east, and south of Upper Klamath Lake. The Klamath River and its tributaries flow through mountainous regions from Iron Gate Dam (Figures 1-2 and 1-3) downstream almost to the coast.

Most of the activities of the Klamath Project occurred in the relatively flat region around Upper Klamath Lake, mainly to the south and east. Much of the region's agriculture occurs in this area, and most of it is below 1,600 m in elevation and depends on irrigation.

The upper Klamath River basin, which includes Upper Klamath Lake is home to 18 species of native fishes, two of which, the shortnose sucker (*Chasmistes brevirostris*) and the Lost River sucker (*Deltistes luxatus*), inhabit the lake and are listed as endangered under the U.S. Endangered Species Act (ESA); one widespread species, the bull trout (*Salvelinus confluentus*), is listed as threatened. The upper basin also is home to 18 species of nonnative fishes, some of which are strains or subspecies of the native species (NRC 2004a).

FIGURE 1-2 Iron Gate Dam on the Klamath River is the dividing point between the upper and lower Klamath River basins. The penstock for the power generators is on the right, and the spillway is on the left in this view looking upstream. SOURCE: Photograph by W.L. Graf, University of South Carolina, July 2006.

FIGURE 1-3 A short reach of the lower Klamath River near Gottsville, California, shows the complexity of the channel and the variety of aquatic habitats in the stream. A shallow bar in the foreground separates a swift-water riffle from a quiet backwater pool on the left.
SOURCE: Photograph by W.L. Graf, University of South Carolina, July 2006.

The lower Klamath River and its tributaries support 20 native fish species, one of which, the coho salmon (*Oncorhynchus kisutch*), is listed as threatened in the basin (and elsewhere in Oregon and California, but not throughout its range). Other anadromous salmonids of interest include Chinook salmon (*O. tshawytscha*) and steelhead, the anadromous form of rainbow trout (*O. mykiss*). All three of the anadromous salmonid species were much more abundant previously than they are today, as described in Chapter 2. Sixteen nonnative species have been reported from the lower Klamath River.

The management and uses of the natural resources of the basin, including water and the fishes, are complex. Many federal, state, county, and other agencies and organizations are involved, and the basin's resources are managed to achieve a variety of divergent purposes.

More information on the region, its biota, human history and human activities there, and management issues are in an earlier NRC report (2004a) and in Chapter 2 of this report.

RECENT HISTORY

We begin our discussion of the recent history of the region, including the events leading to the NRC's involvement, with the very dry year of 2001. The description of the period up to 2002 is adapted from the NRC (2004a) report. The ESA, which pertains in this region to the two endangered suckers of the upper basin and the threatened coho salmon of the lower basin, requires that the U.S. Bureau of Reclamation (USBR) assess the effects of the Klamath Project operations on those species and consult with the U.S. Fish and Wildlife Service (USFWS) about the assessments on suckers and with the National Marine Fisheries Service (NMFS) on coho (USBR 2001a,b). These biological assessments and the USBR's revised assessments in 2002 proposed operations that the USBR judged would offset some of the project's adverse effects on the species (USBR 2002). The USFWS (2001, 2002) endorsed some of the USBR proposals, but concluded that more water was needed to maintain Upper Klamath Lake at levels that would protect the suckers. The NMFS also agreed with some of the URBS proposals, but concluded that more water was needed to maintain higher minimum flows in the Klamath River below Iron Gate Dam than proposed by the USBR (NMFS 2001, 2002). The "biological opinions" of the USFWS and the NMFS indicated that some of the USBR's proposals would jeopardize the continued existence of the listed species, and therefore the USBR was required to allot more water to the lake and to the river than had been planned, leaving less than had previously been allocated for agriculture.

Those restrictive allocations, coupled with a very dry year, resulted in hardships for many of the basin's water users, and the controversy surrounding the allocations became intense. As a result of the controversy, the U.S. Department of the Interior asked the NRC to review the scientific bases of the USBR biological assessments and the USFWS and NMFS biological opinions. In response, the NRC established the Committee on Endangered and Threatened Fishes in the Klamath River Basin, which issued an interim report focused on the biological assessments and biological opinions (NRC 2002), and a final report that took a broader look at strategies for recovery of the endangered and threatened fishes of the basin (NRC 2004a). The interim report concluded that most of the recommendations of the biological opinions had scientific support but that available scientific data did not support the higher minimum lake levels or the higher minimum river flows recommended in the biological opinions to benefit the species listed under the ESA. The later report confirmed those conclusions and included many recommendations for actions to benefit the listed fish species and to improve scientific understanding of the basin.

In addition, a group known as the OSU-UC Davis group (Braunworth et al. 2002) and an independent group of scientists, mainly from the Pacific

Northwest (IMST 2003), also reviewed biological opinions, management, and science in the Klamath basin. The IMST report has a useful chart comparing its conclusions with those of the biological opinions, Braunworth et al. (2002), and the NRC interim report (2002).

Since the publication of the NRC reports in 2002 and 2004, two new documents became available: an estimate of natural or unimpaired flows in the basin as they were before the project was begun (Natural Flow Study or NFS; USBR 2005), and a model of the relationship between flows in the Klamath River and the habitat there for anadromous fishes, especially salmonids and including the threatened coho salmon (*Oncorhynchus kisutch*) (the study often is referred to as Hardy Phase II; here referred to as Instream Flow Study Phase II or IFS) (Hardy et al. 2006a). A more detailed history of these two documents and related ones is in Figure 1-4. Because the new documents have the potential to change scientific conclusions and management options based on earlier information, the Department of the Interior asked the NRC to evaluate them and their implications for the biota of the basin. In response, the NRC established the Committee on Hydrology, Ecology, and Fishes of the Klamath River Basin, which prepared this report.

In addition to the history summarized above, several other developments have occurred in the Klamath basin since the NRC report published in 2004. These developments include the full implementation of the Trinity River restoration program (Schleusner 2006). Scientific advances have occurred since the NRC (2004) report, largely stimulated by the mass mortality of fish in the lower Klamath River of September 2002, when more than 33,000[1] mostly adult fish died in the lower Klamath river, about 95% of which were Chinook salmon, the remainder being mostly steelhead; less than 1% of the deaths were coho salmon. The precise cause or causes of the event cannot be determined (NRC 2004a, CDFG 2004), although the proximate cause was infection with two ubiquitous pathogens, the protozoan *Ichthyopthirius multifilis* and the bacterium *Flavobacter columnare*. The flow and water volume in the river were atypically but not unprecedentedly low, and the water temperatures were high but not exceptionally so. The salmon run was somewhat larger and earlier than average. The CDFG (2004) hypothesized that recent changes in the river channel made upstream migration more difficult during low flows, and thus the fish were concentrated in poor conditions, leading to critically high infections of the pathogens. As the NRC (2004a) recommended, these factors need further investigation.

The advances, described in CDFG (2004) and Hardy et al. (2006a),

---

[1]The California Department of Fish and Game, which made this estimate, described it as "conservative."

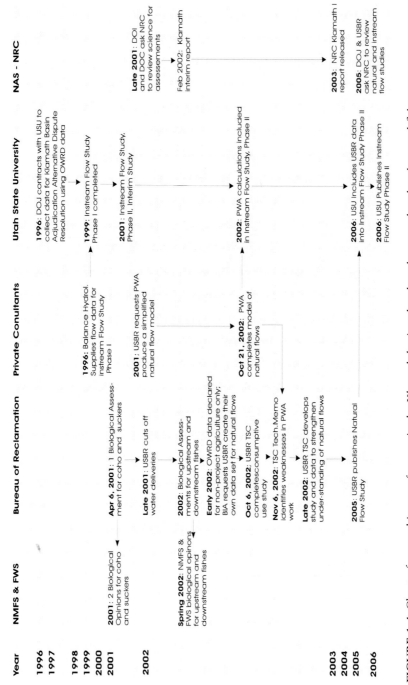

FIGURE 1-4 Chart of recent history of events in the Klamath basin related to threatened and endangered fishes.
SOURCE: Data from published information and from T. Hardy, USU, and J. Hicks, USBR

include an improved understanding of diseases of Klamath River salmon and the factors that cause the diseases to become problematic, new information on thermal refugia and temperatures in the main stem of the Klamath River and its tributaries, and new observations of coho salmon using the main stem as habitat.

THE PRESENT STUDY

Statement of Task

A multidisciplinary committee will be established to evaluate new scientific information that has become available since the National Research Council issued its 2004 report *Endangered and Threatened Fishes in the Klamath River Basin*. The new information to be evaluated by the committee will include two new reports on (1) the hydrology of the Klamath basin and (2) habitat needs for anadromous fish in the Klamath River, including coho salmon. The committee will also identify additional information needed to better understand the basin ecosystem.

To complete its charge, the committee will

1. Review and evaluate the methods and approach used in the Natural Flow Study to create a representative estimate of historical flows and the Hardy Phase II studies, to predict flow needs for coho and other anadromous fishes.
2. Review and evaluate the implications of those studies' conclusions within the historical and current hydrology of the upper basin; for the biology of the listed species; and separately for other anadromous fishes.
3. Identify gaps in the knowledge and in the available scientific information.

The Committee's Process

To execute its charge, the committee met four times: in Sacramento, CA. February 13-14, 2006; in Yreka, CA. October 2-4, 2006; in Klamath Falls, OR. January 29-31, 2007; and in Irvine, CA. April 11-13, 2007. At the first three meetings, the committee heard presentations from scientists and others, including agency officials, familiar with various aspects of the region and the operation of the Klamath Project; the committee also received presentations from the public (see list of presenters in Appendix B). At its second meeting, the committee visited a restoration and research project on the upper Shasta River, the Iron Gate Dam and hatchery on the Klamath River, and the monitoring station near the mouth of the Shasta River. Individual members of the committee and staff also visited other parts of the basin, including Upper Klamath Lake; the Williamson, Sprague, and Wood rivers; the Link River and Link River Dam; and Keno Dam. Groups of committee and staff members visited Dr. Thomas Hardy in Logan, Utah, on October 1, 2006 and the USBR office in Denver, Colorado, on November 20, 2006, to discuss the respective reports with their authors in detail.

Relationship of This Report to Previous NRC Reports

This is the third NRC report on the Klamath River basin and its fishes. The first (NRC 2002) focused narrowly on the scientific bases for the biological opinions of the USFWS and the NMFS and the biological assessments of the USBR. The second (NRC 2004a) took a broad look at the Klamath basin and considered options for reversing the declines of the listed species of fishes. The present report was requested after two significant documents were made public (USBR 2005, Hardy et al. 2006a), and it addresses the documents in some detail. However, this report also addresses the implications of the two reports for the anadromous fishes in the Klamath River and the broader context in which science is conducted in the basin. New developments have occurred since the previous reports were published, and this report is not a revisiting of the issues covered by the earlier NRC reports. Indeed, this committee endorses the recommendations of the earlier reports for reversing the declines of the listed species, and this report should be considered as building on the previous ones and continuing where they left off.

REPORT ORGANIZATION

Chapter 2 provides a description of the Klamath basin, along with descriptions of its hydrology and biota. There is a description of the life histories of three anadromous species of greatest interest: coho salmon, Chinook salmon, and steelhead. Chapter 3 provides an analysis of the use and development of models, as well as their capabilities and shortcomings. The considerable detail of this chapter is important because models are central to the two documents this committee reviewed; therefore, the appropriate context is required for evaluating them. Chapters 4 and 5 provide descriptions, analyses, and evaluations of the Natural Flow Study and the Instream Flow Study Phases I and II, respectively. Chapter 6 presents a discussion of systematic approaches to the use of science in decision making and their relevance to the scientific activities that have been and are being conducted in the Klamath River basin. Chapter 7 presents the committee's conclusions and recommendations.

2

The Klamath Basin

The Klamath basin has long been celebrated for its lakes, streams, forests, hunting, fishing, and agriculture. In particular, the Klamath River was once the third-largest salmon-producing stream on the West Coast behind the Sacramento and Columbia Rivers (EPA 2006). This chapter provides a brief summary of the social, economic, and biological resources of the basin. Further detail can be found in the NRC (2004a) report on the Klamath River. To set the following chapters in their physical and biological contexts, this chapter provides a broad regional introduction to the Klamath River basin by describing its physical geography, geology, and hydrology. The chapter continues with a description of the fish communities of the basin and a brief review of the human institutions that manage these physical and biological resources. Finally, the chapter summarizes the changes in physical and biological systems brought about by their human management.

DESCRIPTION OF THE BASIN

Physical Characteristics and Land Use

The Klamath basin is located in south-central Oregon and northwestern California (Figure 1-1). The basin drains approximately 16,000 mi^2 with 35% of the watershed in Oregon and 65% in California. The uppermost reaches of the watershed originate in Oregon, and the main-stem river flows through the basin for about 250 miles and enters the Pacific Ocean about 20 miles south of Crescent City, CA, in Del Norte County. In Oregon, the basin occupies portions of Jackson, Lake, and Klamath counties; in California, it flows through the counties of Siskiyou, Modoc, Trinity, Humboldt, and Del Norte (NRCS 2006).

For discussion and management, the Klamath basin is divided into the upper and lower Klamath basins. The generally accepted boundary between the two is Iron Gate Dam on the Klamath River. All lands upstream (that is, east and west) of the dam are within the upper Klamath basin (area: 8,060 mi^2); and those below (that is, south and west of) the dam comprise the lower Klamath basin (area: 7,628 mi^2). The lands within the upper basin fall within the U.S. Bureau of Reclamation (USBR) Klamath Project. A number of sub-basins are present throughout the watershed (Table 2-1). A large portion of the upper basin is in agriculture and rangeland use, whereas in the lower basin, forest land dominates the landscape with the exception of the Scott and Shasta basins which have large portions of area in agriculture and rangeland (Figure 2-1).

The largest towns in the basin are Klamath Falls, Oregon, which has a total metropolitan population of about 42,000 (City of Klamath Falls 2007); Yreka, CA (7,300), (Yreka Chamber of Commerce 2007); and Weaverville, California (3,550) (City-Data.com 2007). The basin is home to six federally recognized American Indian tribes: Yurok, Hoopa Valley, Karuk, Quartz Valley, Resighini (all in California) and the Klamath in Oregon.

TABLE 2-1 Klamath Basin Sub-basins Shown in Figure 2-1

USGS Identification	Name	Area in acres
	Upper Klamath Basin	
18010201	Williamson River	934,490
18010202	Sprague River	1,029,824
18010203	Upper Klamath Lake	464,903
18010204	Lost River	1,926,303
18010205	Butte Creek	386,034
18010206E	Upper Klamath, East Section	416,786
Upper Klamath Basin Total (acres; sq mi)		5,158,340; 8,060
	Lower Klamath Basin	
18010206W	Upper Klamath, West Section	489,887
18010207	Shasta River	508,841
18010208	Scott River	520,612
18010209	Lower Klamath River	984,709
18010210	Salmon River	480,178
18010211	Trinity River	1,303,253
18010212	South Fork Trinity River	594,895
Lower Klamath Basin Total (acres; sq mi)		4,882,015; 7,628

SOURCE: Modified from NRCS 2006.

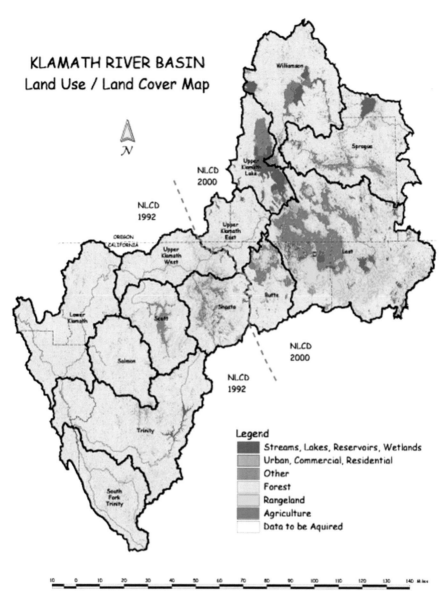

KLAMATH RIVER BASIN
Land Use / Land Cover Map

FIGURE 2-1 Land use throughout the Klamath basin divided into sub-basins.
SOURCE: NRCS, National Hydrography data set, December 4, 2002.

The upper and lower basin economies are similar in size and output; however, the products are very different. The economy of the upper Klamath basin, which is home to approximately 120,000 people, is heavily dependent upon agriculture, the forest-products industry, tourism, and public employment. In 1998, the area had 60,000 jobs, produced about $4 billion in output, and added about $2.3 billion in value to purchased inputs (NRC 2004a). The lower Klamath basin, which is home to about 167,000 people, in 1998 produced $5.9 billion of output, added more than $3.3 billion in value to purchased inputs and had more than 84,000 jobs. The greatest numbers of jobs were represented by the retail trade, educational services, and health care and social assistance industries (NRC 2004a).

The prominent water feature of the Klamath basin is Upper Klamath Lake, the largest lake in Oregon (Oregon Lakes Association 2005). Upper Klamath Lake varies in width from about 6 to 14 miles and is about 25 miles long (USBR 2005). Upper Klamath Lake's perimeter is about 88 miles, its surface area is about 61,520 acres (96 mi^2), the mean surface elevation is about 4,140 feet above mean sea level, the mean depth about 13 feet, and the maximum depth about 49 feet (Oregon Lakes Association 2005). The USBR maintains the lake's surface elevation at 4,136 to 4,146 feet above mean sea level by virtue of a dam constructed in 1917 (Oregon Lakes Association 2005). The USBR estimates the lake's total capacity to be about 650,000 acre-ft with an operational capacity of about 486,800 acre-ft; its net mean annual inflow is 1,200,000 acre-ft, ranging from 576,000 to 2,400,000 acre-ft (USBR 2005). As an important component of water-resource utilization in the region, Klamath Lake provides water for irrigation and power generation. Other lakes in the upper Klamath basin include Lower Klamath Lake (4,700 acres; 7.3 mi^2); Tule Lake (9,450-13,000 acres; 14.8-20.3 mi^2); Clear Lake (highly variable area; average of 21,000 acres; 32.8 mi^2); and Gerber Reservoir (highly variable area) (Figure 2-1).

The upper basin has several tributaries. The Williamson and Wood rivers provide the major flow contribution to Upper Klamath Lake. The Sprague River is a tributary to the Williamson River, and Chiloquin Dam, which is slated for removal, is located just upstream of the confluence with the Williamson River. The Sycan River is a major tributary of the Sprague River. Link River flows from Klamath Lake into Lake Ewauna, from which the Klamath River emanates.

There are six main-stem dams in the upper Klamath basin, listed in order downstream from Upper Klamath Lake: Link River, Keno, J.C. Boyle, Copco No. 1, Copco No. 2, and Iron Gate. Link River Dam is for irrigation purposes and for controlling lake levels in Upper Klamath Lake (Figure 2-2a-d). All other dams except for Keno generate power. Reaches of the Klamath River below J.C. Boyle Dam experience substantial daily fluctuations in response to operating rules for dams that generate electrical power to meet peak demand periods, while flows above this structure have flows that change more gradually.

Relicensing hearings by the Federal Energy Regulatory Commission

FIGURE 2-2a This reach of the Link River below Upper Klamath Lake, shown here in 1919, is the site of Link River Dam. Note the bedrock outcrops forming ledges and rapids in the channel that act as a sill for the level of the lake upstream.
SOURCE: Boyle 1976. Reprinted with permission; copyright 1976, Klamath County Museum.

FIGURE 2-2b The newly completed Link River Dam spans the channel and diverts much of the river's discharge into the Keno Canal on the right bank in this 1922 image made from the same location as in Figure 2-2a.
SOURCE: Boyle 1976. Reprinted with permission; copyright 1976, Klamath County Museum.

(FERC) are under way at this writing. The six dams block access of migratory fish to their historical upstream spawning habitat, and their removal has been proposed as a potential option for fishery restoration. Current is-

FIGURE 2-2c The Link River Dam, shown from the same location as in Figures 2-2 a and b, diverts water into the Keno Canal and the Ankeny Canal in this view made about 1940. Riparian vegetation has colonized channel areas downstream from the dam. (Boyle indicated that the date of the image is 1924, but given the size of the trees in the image that were completely absent only 2 years previously, as shown in Figure 2-3b, 1924 is highly unlikely.)
SOURCE: Boyle 1976. Reprinted with permission; copyright 1976, Klamath County Museum.

FIGURE 2-2d The Link River Dam, shown in this 2006 view from the same location as the views in Figures 2-2 a, b, and c, now includes a fish ladder, recently installed near the right abutment (left side of the dam in this image).
SOURCE: Photograph by W.L. Graf, University of South Carolina, July 2006.

sues regarding hydropower include the fact that the Klamath dams produce less than 1% of the energy demanded by Pacificorp's customers and that low electric rates for the Klamath Reclamation area have resulted in little incentive to conserve water by reducing pumping of irrigation water.

The Klamath-Trinity river system is the largest between the Sacramento and Columbia rivers in terms of flow, salmon production, and economic importance, and one of the most highly modified. In the lower basin, there are four major tributaries: the Shasta River, Scott River, Salmon River, and the Trinity River. Many smaller tributaries enter the Klamath River between Iron Gate and Orleans. The Trinity River watershed, draining 2,966 mi^2, is the largest tributary watershed to the Klamath River and comprises about 19% of the total basin area. With an average annual precipitation of about 57 in., the watershed produces more runoff and sediment than any other Klamath River tributary. The narrow alluvial corridors on the main-stem Trinity River and its largest tributary, the South Fork of the Trinity River, allow for a meandering stream with coarse-grained channels that provide excellent spawning and rearing grounds for coho salmon and other anadromous fishes.

The size of the watershed and its high-quality spawning and rearing grounds made the Trinity River extremely productive for anadromous fishes (USFWS/HVT 1999). The smaller Salmon River watershed encompasses about 750 mi^2. Its lack of large alluvial valleys means that the land-use practices that can severely affect anadromous fishes are limited, thereby enhancing its fishery characteristics as opposed to the Scott and Shasta watersheds (NRC 2004a).

Water runoff from precipitation events is buffered by the landscape in the upper watershed, and thus runoff production in the entire basin is heavily weighted toward the lower basin watersheds. Despite the fact that it comprises more than 50% of the entire basin, the upper Klamath basin produces only about 12% of the average annual runoff, which is approximately 13 million acre-ft at the mouth of the Klamath River (NRC 2004a). The upper Klamath basin produces less runoff as a result of the generally low relief, presence of marshes and wetlands (which increase hydraulic residence times), and its location in the rain shadow of the Cascades (NRC 2004a). In contrast, the lower Klamath basin watersheds, near the coast, have portions with as much as 100 in. of annual precipitation. In addition, elevations above 5,000 feet often have winter and spring snowpacks in wet years, which produce much runoff during warm winter storms (NRC 2004a).

The physiography of the Klamath basin is quite different from most watersheds in that the greatest relief and topographic complexity occur in the lower basin (Mount 1995). The Klamath basin occurs at or near the convergence of several tectonic plates: the Pacific, Juan de Fuca, and North

American plates (see Figure 2-1 from NRC 2004a), the subducting Juan de Fuca plate off the northern California and Oregon coasts (and farther north, the Gorda plate), moving beneath the North American plate, has given rise to the Cascade Mountains, a volcanic arc. Two of the more prominent Cascade volcanoes, Shasta and Mazama (in whose caldera Crater Lake now sits), are in this volcanic arc. Most of the upper Klamath basin lies within the so-called back-arc area, whereas the lower Klamath basin lies within the dynamic fore-arc area. The volcanic arc essentially separates these two basins. The rapid tectonic uplift of portions of the lower Klamath basin is evidenced by the steep, rugged watersheds of the Salmon and Trinity rivers. The other major lower Klamath watersheds, the Scott and the Shasta, have broad, alluvial valleys in the central portions that support agriculture.

Fish Communities in the Klamath River Basin

Lower Basin Fishes.   The Klamath basin below Iron Gate Dam supports a fish community mainly comprising anadromous fishes that spend a portion of their lives in fresh water and a portion in the ocean. There are 19 species of native fishes in the Klamath basin below Iron Gate (Table 2-2). Of the 19 species, 13 are anadromous and 2 are amphidromous (larval stages in saltwater). In addition to these 19 species, another 17 nonnative species are present in the lower basin, of which only two (American shad [*Alosa sapidissima*] and occasionally brown trout [*Salmo trutta*]) are anadromous. The nonnative species are mainly warm- and cool-water species that thrive in slow-current or reservoir environments (NRC 2004a).

Anadromous Species.   Species with a life history of anadromy reproduce or spawn in freshwater rivers or lakes; the young then migrate out to sea, or "smolt," to grow to maturity and return to their natal streams. The strategy of anadromy is thought to have evolved as an approach to take advantage of the relatively protected environments found in rivers while exploiting the productive abundance of the ocean to grow to large sizes (Gross 1987). The process of smoltification in the emigrating juvenile fish is physiologically complex; it prepares the young fish for life in saltwater conditions (McCormick and Saunders 1987). The process results in both internal and external physical changes as well as in behavioral changes. The young fish become more slender and elongated, their internal organs prepare for life in saltwater, and the fish school and move downstream together. This process is cued and supported by photoperiod and river temperatures and flows. A narrow window of water temperature, which is species specific, supports the smoltification process in early spring. Increases in stream temperature above the smolting thresholds will result in juvenile fish delaying emigra-

TABLE 2-2 Native Fishes of the Lower Klamath River and Its Tributaries and Their Status

Name	Status	Comments
	Anadromous Species	
Chinook salmon, *Oncorhynchus tshawytscha*		Much reduced in numbers
Southern Oregon-Northern California ESU[a]		1. Much reduced in numbers; focus of hatchery supplementation
Upper Klamath and Trinity Rivers ESU		
1. Fall run		2. Possibly extinct
2. Late fall run		3. Endangered but not recognized as an ESU; distinct life history
3. Spring run		
Chum salmon, *O. keta*	Rare; state special concern	Southernmost run of species; TTS[b]
Coastal cutthroat trout, *O. clarki clarki*	State, special concern	Found only in lower main stem and tributaries; resident populations above barriers; TTS
Coho salmon, *O. kisutch*; Southern Oregon–Northern California ESU	Federally threatened	Being considered for state listing; TTS
Eulachon, *Thaleichthys pacificus*	State, special concern	TTS; huge runs were once present in lower 5-7 miles of Klamath River
Green sturgeon, *Acipenser medirostris*	State, special concern; posed for listing	TTS; important fishery; infrequently observed as far upstream as Iron Gate Dam
Longfin smelt, *Spirinchus thaleichthys*	State, special concern	Small population mainly in the estuary
Pacific lamprey, *Lampetra tridentate*	Declining	TTS; probably multiple runs
Pink salmon, *O. gorbuscha*	Extirpated	TTS; infrequent captures do occur (Hardy et al. 2006)
River lamprey, *L. ayersi*	Uncommon	Little known
Steelhead (rainbow trout), *O. mykiss*	Common but declining, posed for listing	TTS; nonmigratory populations present above barriers
Klamath Mountains Province ESU	Most abundant	
Winter run	Endangered but not listed as a separate ESU	
Summer run		

continued

TABLE 2-2 Continued

Name	Status	Comments
Threespine stickleback, *Gasterosteus aculeatus*	Common	Near the ocean exhibits anadromy; farther upstream present as nonmigratory
White sturgeon, *A. transmontanus*	Uncommon	May not spawn in the river; infrequently observed as far upstream as Iron Gate Dam
Amphidromous[c]		
Coastrange sculpin, *C. aleuticus*	Common	Larvae wash into estuary
Prickly sculpin, *Cottus asper*	Common	Larvae wash into estuary
Nonmigratory		
Klamath River lamprey, *L. simulis*	Common	Little known
Klamath small-scale sucker, *Catostomus rimiculus*	Common, widespread	
Klamath speckled dace, *Rhinichthys osculus*	Common, widespread	
Lower Klamath marbled sculpin, *C. klamathensis polyporus*	Common	Endemic

[a]Evolutionarily significant unit.

[b]Tribal trust species.

[c]*Amphidromous* species, sometime called euryhaline species, can move back and forth between fresh, brackish, and salt water at various life stages, but they do not normally do so for breeding, as anadromous species do.

SOURCE: Adapted from NRC 2004a.

tion and possibly remaining in a stream environment unsuitable for their survival.

The duration spent in either the ocean or river habitat is specific to each species and strains within species, as is the timing of emigration of juveniles and immigration of adults. The historical hydrologic conditions of the river shaped and defined the instream habitats that supported these various life history strategies. Examples include the development of clean gravels in riffles for invertebrate production and juvenile feeding habitat, as well as the high flow scouring action to maintain deeper pools for adults holding in the river before spawning. Natural variability of climatological conditions in the basin produced some years that were more favorable for one species than for others. Thus the development of these various life histories within species is an important strategy for ensuring reproductive success through a variety of habitat conditions. Physical instream habitat in many areas throughout the basin has been altered as a result of land-use activities that increased sedimentation, channelization, decreased streamside riparian vegetation, and physical blockage of fish from upstream habitat areas. Alteration of the flow regime in the basin as a result of hydropower, irrigation, and other off-channel uses has resulted in changes in the annual pattern of stream flow as well as altered the thermal properties of the river and its tributaries. In some areas, the combination of altered flow regime and land-use activities has resulted in water quality conditions that do not support all of the life history phases of the salmonids that use the area. Understanding the complexity of the life history needs of the anadromous fish community, reflection on the current status of the stocks, and an overview of the instream habitat conditions are important for defining and evaluating the management directions and goals for the basin.

Of the anadromous species complex, much of the fishery focus is on those species of high commercial and recreational value, such as Chinook salmon, coho salmon, and steelhead. Other anadromous fishes of general interest in the lower Klamath River include tribal trust species, such as the green sturgeon, white sturgeon, Pacific lamprey, eulachon, and some other species, but these fishes have had less economic importance than the three salmonids, and they are not federally listed. Most of them are more common in the lowest part of the river than farther upstream, and they have not been the subject of as many studies or the focus of as many recent controversies as the salmonids. For these reasons, they receive less attention than the salmonids in this report. However, the relative scarcity of information on them, noted also by Hardy et al. (2006a), is an impediment to developing management plans for all the anadromous species, rather than for coho, Chinook, and steelhead (see Table 2-3).

Before development of the basin, anadromous species ranged widely through the tributaries and upstream of Upper Klamath Lake into the Sprague and Williamson basins and Spencer Creek (Coots 1962; Fortune et al. 1966; Hamilton et al. 2005, as cited in Hardy et al. 2006a). Access to the upper Klamath basin by anadromous species ended with completion of Copco Dam No. 1 in 1918, the reduction in access occurring during earlier construction of the Lost River diversion canal and Chiloquin Dam in 1912-1914 (Hardy et al. 2006a).

Upper Basin Fishes.   This report focuses more on the lower than the upper basin fishes, but for completeness, Table 2-4 describes the native and nonnative fishes of the upper Klamath River basin.

TABLE 2-3 Chinook Salmon, Coho Salmon, and Steelhead Life Periodicities Between Iron Gate Dam and the Trinity River

	Oct.	Nov.	Dec.	Jan.	Feb.	Mar.	Apr.	May	Jun.	Jul.	Aug.	Sep.
Iron Gate Dam to Shasta River												
Chinook salmon fry									1			
Chinook salmon juvenile												
Chinook salmon spawning/incubation												
Coho salmon fry												
Coho salmon juvenile												
Steelhead fry												
Steelhead spring juvenile												
Steelhead summer juvenile												
Steelhead juvenile (all)												
Scott River to Trinity River												
Chinook salmon fry												
Chinook salmon juvenile												
Chinook salmon spawning/incubation												
Coho salmon fry												
Coho salmon juvenile												
Steelhead fry											2	
Steelhead spring juvenile												
Steelhead summer juvenile												
Steelhead juvenile (all)												

NOTE: Light shading, unless otherwise noted in 1 and 2, indicates occasional use.

1 = Only occasional use from Iron Gate Dam to Shasta River; more frequent use from Shasta River to Scott River.
2 = Only occasional use from Scott River to Salmon River; more frequent use from Salmon River to Trinity River.

SOURCE: Adapted from Hardy et al. 2006a. Reprinted with permission; copyright 2006, Utah State University.

TABLE 2-4 Native and Nonnative Fishes of the Upper Klamath Basin and Their Status

Name	Status	Comments
Native Fishes of Upper Klamath Basin		
Pacific lamprey, *Lampetra tridentata*	Common	TTS[a]; Found in cold-water creeks; isolated from downstream populations; anadromous forms probably spawned as far upstream as Spencer Creek
Klamath River lamprey, *L. similis*	Common	Little known
Miller Lake lamprey, *L. milleri*	Uncommon	Once thought extinct
Pit-Klamath brook lamprey, *L. lethophaga*	Assumed common	Shared with Pit River drainage
Minnows – Cyprinidae		
Klamath tui chub, *Siphatales bicolor bicolor*	Abundant	Widespread in upper Klamath basin
Blue chub, *Gila coerulea*	Common, state special concern (CA)	
Klamath speckled dace, *Rhinichthys osculus klamathensis*	Assumed common	
Suckers – Catostomidae		
Shortnose sucker, *Chasmistes brevirostris*	Federally endangered	
Lost River sucker, *Deltistes luxatus*	Federally endangered	
Klamath largescale sucker, *Catostomus snyderi*	Assumed common	
Klamath smallscale sucker, *C. rimiculus*	Uncommon	Common in lower Klamath basin
Salmon and Trout – Salmonidae		
Klamath redband trout, *Oncorhynchus mykiss* subsp.	Common	
Coastal steelhead, *O. mykiss irideus*	Extirpated	TTS; historically occurred in tributaries of Upper Klamath Lake; common but declining in lower basin

continued

TABLE 2-4 Continued

Name	Status	Comments
Coho salmon, *O. kisutch*	Extirpated	TTS; historically occurred as far upstream as Spencer Creek; still present in small numbers in lower basin
Chinook salmon, *O. tshawytscha*	Extirpated	TTS; historically occurred in tributaries of Upper Klamath Lake, particularly the Williamson and Sprague Rivers; some runs still common in lower basin
Bull trout, *Salvelinus confluentus*	Federally threatened	Restricted to 10 streams in basin with temperatures <18°C
Sculpins – Cottidae		
Upper Klamath marbled sculpin, *Cottus klamathensis klamathensis*	Common	Widespread in basin
Klamath Lake sculpin, *C. princeps*	Common	Abundant in Upper Klamath Lake
Slender sculpin, *C. tenuis*	Uncommon	Extirpated from much of former range; currently found in lower Williamson River and Upper Klamath Lake

Nonnative Fishes

Name	Status	Comments
Minnows – Cyprinidae		
Goldfish, *Carassius auratus*	Uncommon	Exotic[b]; locally common in pond
Golden shiner, *Notemigonus chrysoleucas*	Uncommon	Introduced[c]; baitfish
Fathead minnow, *Pimephales promelas*	Common	Introduced; widespread in basin; often the most abundant fish species
Catfishes – Ictaluridae		
Brown bullhead, *Amieurus nebulosus*	Common	Introduced; increasing range
Black bullhead, *A. melas*	Uncommon	Introduced
Channel catfish, *Ictalurus punctatus*	Unknown	Introduced; may not be established
Salmon and Trout – Salmonidae		
Kokanee salmon, *O. nerka*	Uncommon	Introduced
Rainbow trout, *O. mykiss*	Common	Introduced; various hatchery strains are widely planted; poor survival because of endemic diseases

TABLE 2-4 Continued

Name	Status	Comments
Brown trout, *Salmo trutta*	Common	Exotic; anadromous and resident forms also found in lower basin
Brook trout, *Salvelinus fontinalis*	Uncommon	Introduced; can hybridize with and otherwise displace native bull trout
Sunfishes – Centrarchidae		
Sacramento perch, *Archoplites interruptus*	Common	Introduced; increasing in range
White crappie, *Pomoxis annularis*	Uncommon	Introduced; mainly in a few reservoirs
Black crappie, *P. nigromaculatus*	Uncommon	Introduced; recorded from Lost River
Green sunfish, *Lepomis cyanellus*	Common	Introduced; widespread in reservoirs, hybridizes readily with other *Lepomis* spp.
Bluegill, *L. macrochirus*	Uncommon	Introduced; locally abundant in ponds/reservoirs
Pumpkinseed, *L. gibbosus*	Common	Introduced; widespread
Largemouth bass, *Micropterus salmoides*	Common	Introduced; common in reservoirs
Perch – Percidae		
Yellow perch, *Perca flavescens*	Common	Introduced; common in large reservoirs

[a]Tribal trust species.
[b]*Exotic* means introduced from outside the United States.
[c]*Introduced* means introduced to the basin from elsewhere in the United States.

SOURCE: Adapted from Moyle 2002; NRC 2004a; Hamilton et al. 2005.

Management Institutions

Multiple management authorities with different roles and responsibilities are present throughout the basin (see Table 2-5). Complications arise as many of the mandates for the agencies are often seemingly in direct conflict with one another. At no time was this conflict more acute than in 2001 when irrigation water was shut off for the water users and retained for the purposes of managing for endangered fish species (NRC 2004a).

Basin groups exist often as a result of legislation to bring together social, political, and biological interests for compromise and management of the basin's resources. Sub-basin groups arise around specific interests, such

TABLE 2-5 Species Life Stages and Concerns for Management As Noted by Month

Month	Salmonid Species-Specific Life History Concerns	Notes
January	All main-stem anadromous spawners incubation	Ensure flow recommendations
February	Chinook and coho salmon fry, juvenile coho salmon and steelhead rearing, and half-pounders	Second half of month reflects flow needs for swim-up Chinook salmon fry, mid-february beginning of 0+ and 1+ coho smolt outmigration, juvenile steelhead rearing
March	Chinook and coho salmon fry and presmolts, steelhead rearing, beginning of steelhead smolt outmigration and half pounders	Priority to create edge-water habitat for Chinook and coho salmon fry, juvenile steelhead and coho presmolts habitat
April	Coho and Chinook salmon fry rearing, Chinook and coho salmon smolt, steelhead juvenile rearing, and steelhead smolt outmigration	Flows to enhance rearing and reduce transit times. Consider temperature modeling in flow recommendations to offset water temperatures that enhance C. Shasta transmittal; mid-April begin peak 1+ coho outmigration; coho and Chinook salmon fry habitat
May	Chinook and coho salmon fry rearing and smolt outmigration, hatchery Chinook salmon release, steelhead juvenile rearing and smolt outmigration	Flows to enhance rearing and reduce transit times; continue peak 1+ coho outmigration; spring Chinook salmon adult immigration; Chinook and coho salmon fry; disease considerations
June	Coho salmon 0+ rearing and 1+ outmigration, hatchery Chinook salmon release in all reaches, steelhead rearing and smolt outmigration	Late June end 1+ coho salmon outmigration; coho salmon 0+ rearing; spring Chinook salmon adult migration; hatchery competition; disease considerations
July	Juvenile Chinook and coho salmon and steelhead	Consider a floor flow for drier exceedances as smolts need to get to estuary; hatchery competition; disease considerations
August	August 1-15 all juveniles, August 16-31 adult Chinook salmon passage and staging habitat	Flows reflecting July recommendations for first half of month
September	Adult Chinook salmon passage, coho and Chinook salmon and steelhead rearing, adult steelhead and half-pounders	Disease considerations; facilitate passage of adults
October	Adult Chinook salmon main-stem spawning, juvenile steelhead, Chinook and coho salmon rearing, adult steelhead and half pounders	

TABLE 2-5 Continued

Month	Salmonid Species-Specific Life History Concerns	Notes
November	Adult coho and Chinook salmon main-stem spawning, all juveniles rearing, adult steelhead and half pounders	Consider flows to inundate key side channels for overwintering habitat; maintain flows to reduce dewatering of redds
December	Coho spawning, all main-stem anadromous spawner incubation, all juvenile rearing, and half pounders	Maintain flows to reduce dewatering of redds; provide juvenile habitat

SOURCE: Adapted from Hardy et al. 2006a. Reprinted with permission; copyright 2006, Utah State University.

as fishery management or specific watershed improvements. Stakeholder groups are special interest groups, often with a narrow focus pertaining to one or more specific issue such as irrigation or salmon rehabilitation. Some groups arise as a reactionary response to a crisis. Within the Klamath basin, 5 basin groups, 17 sub-basin groups, 21 stakeholder groups, 6 federal agencies, 7 tribal agencies, 2 states, 9 state agencies, 7 counties, and 9 local municipalities are somehow involved in the management interests of the Klamath basin (Table 2-6). These groups have varying levels of activity and involvement and a website (www.onebasin.org) provides access to information on each of these organizations for the purposes of communicating basin activities and initiatives. In addition, the Klamath Settlement Group, consisting of 26 stakeholder groups, is working on an agreement scheduled for November 2007 concerning dam operations in the Klamath River and "ideas to restore the fisheries, meet the economic needs of irrigators, tribes and local governments and protect water quality and agriculture" (Fletcher and Addington 2007). The particulars and discussions are confidential, so no additional information is available at the time of this writing.

HUMAN-INDUCED CHANGES IN THE BASIN

Changes in Flows, Sediments, and Channels

The Klamath River has been profoundly altered by human settlement and resource exploitation. Hydrologic alterations include changes in runoff from timber harvest, other changes in vegetative cover and land use, diversions and storage for agriculture and hydroelectric production, diking off of formerly flooded lands, and cutoff of historical flood overflow into Lower Klamath Lake. Geomorphic alterations are multiple; some are hardrock

TABLE 2-6 A Sampling of the Many Government Agencies, Stakeholder Organizations, and Working Groups in the Klamath Basin

Basin Groups

Klamath Basin Coordinating Group
Klamath River Basin Fisheries Task Force
Trinity River Restoration Task Force

Klamath Basin Compact Commission
Upper Klamath Basin Working Group

Sub-basin Groups

Upper Basin
Upper Klamath Watershed Council
Klamath River Working Group
Sprague River Working Group
West Klamath Lake Working Group
Cloverleaf Stewardship Group
Upper Williamson Catchment Group
Lower Williamson Working Group
Klamath Project Area Working Group
Urban Issues Working Group

Mid Basin
Mid-Klamath Watershed Council
Shasta River Coordinated Resources
   Management & Planning
Scott River Watershed Council
Salmon River Restoration Council

Lower Basin
Klamath Fishery Management Council
Trinity River Adaptive Management Work
   Group
Trinity River Fisheries Task Force
Trinity Management Council

Stakeholder Groups

California Trout
Friends of the Trinity River
Klamath Basin Crisis
Klamath Basin Haygrowers Association
Klamath Bucket Brigade
Klamath Outdoor Science School
Klamath Salmon Action Network
Nature Conservancy
Oregon Waterwatch
Siskyou County Farm Bureau
Pacific Coast Federation of Fishermen's
   Associations

Klamath Basin Coalition
Klamath Basin Ecosystem Foundation
Klamath Basin Rangeland Trust
Klamath Forest Alliance
Klamath Restoration Council
Klamath Water Users Association
Oregon Trout
Oregon Wild
RCAA Natural Resources Services
Water for Life

Government Agencies

Federal
Bureau of Land Management
Bureau of Reclamation
Fish and Wildlife Service
Forest Service
National Marine Fisheries Service
Natural Resources Conservation Service

State
Oregon
Water Resources Department
Watershed Enhancement Board
Department of Fish and Wildlife
Department of Environmental Quality

TABLE 2-6 Continued

Government Agencies (continued)		
Tribal	*Counties*	
Hoopa Tribe	Del Norte	
Karuk Tribe of California	Humboldt	
Klamath Tribes	Klamath Lake	
Quartz Valley Indian Community	Modoc	
Yurok Tribe of California	Siskyou	
	Trinity	
Intertribal		
Fish and Water Commission	*Cities*	
Klamath Basin Tribal Water Quality Workgroup	Bonanza	Klamath Falls
	Chiloquin	Malin
	Dorris	Merrill
	Tulelake	Weaverville
	Yreka	
	California	
	Biodiversity Council	
	Department of Fish and Game	
	Department of Water Resources	
	Five Counties Salmonid Conservation Program	
	Ocean Protection Council	
	Water Resources Control Board	
	National Coast Regional Water Quality Control Board	

SOURCE: Modified from information at www.onebasin.org.

and placer mining, using the river to float logs downstream to sawmills and building splash dams to release large volumes of water abruptly to carry logs downstream in a wave, and blasting rock outcrops in the bed of the river to improve log passage. The timber harvest and transport in the upstream, volcanic part of the basin is well documented, including the log drives to the large mill at Klamathon, a now-abandoned site several miles upstream of Hornbrook (Shaw Historical Library 1999, 2002; Beckham and Canaday 2006). Those activities probably had the effect of simplifying channel form through the direct elimination of bedrock and other channel irregularities that interfered with the efficient flow of water and the physical effect of the logs themselves battering the banks.

Mining occurred downstream of Hornbrook, along the axis of Cottonwood Creek, where there is a sharp contact between the volcanic Cascades and the Klamath geologic provinces. The Klamath Province includes a wide range of rock types, including ores of gold and other precious metals. Numerous mining claims (that sought to follow mineralized veins) are visible on hillslopes, and accumulation of gold in alluvial deposits led to extensive placer mining along the river in the nineteenth and early twentieth centuries. Most of the alluvial bottoms of the river downstream of Hornbrook were reworked by placer mines, often from valley wall to valley wall (Ayres Associates 1999). Such reworking would include displacing the channel and in excavating down to bedrock, piling gravel into linear tailing deposits. It probably resulted in increased exposure of the bedrock area and in hyporheic exchange (exchange between shallow groundwater and surface water beneath and adjacent to the streambed). Hardrock mining in the uplands draining to the river would have increased delivery of fine (and some angular coarse) sediments to the channel. Dredging of gravels on the flood plain would have simplified the channel through direct modification and, in many cases, displacement to the other side of the valley so that gravels below the current channel could be mined. Dredging and processing of the placer deposits would have released fine sediments into the water column, potentially damaging aquatic habitat.

Dams and diversions have had geomorphic effects, although the effects are less striking than they are in rivers where the dams are larger and impound flows with greater sediment loads, and where the downstream channels are fully alluvial. The dams on the main-stem Klamath are located in the upper river, within the volcanic lithologies of the basin and Range Province (which includes the Cascade and Modoc geologic provinces), upstream of the Cottonwood Creek confluence near Hornbrook. This part of the basin has less rainfall, lower sediment yields, and more bedrock-controlled channel (and thus less alluvial channel) than the Klamath Province downstream. As described elsewhere, six dams are part of the Klamath River hydroelectric project owned by PacifiCorp (the "PacifiCorp Project"): Link River, Keno, J.C. Boyle, Copco 1 and 2, and Iron Gate (Boyle 1976, PacifiCorp 2004) (Figure 2-3).

In considering how these dams might affect flow, channel form, sediment, and ultimately habitat on the Klamath River, it may be helpful to review the effects observed generally in the literature. Recent NRC reports also have reviewed the effects of dams on salmon (for example, NRC 1996, 2004b). Much of the concern about dam effects on fish habitat has been about salmonid spawning gravels. Dams can affect spawning gravels in two principal ways. When reservoirs are large enough to reduce floods, fine sediment from tributaries (and from bank erosion and other sources) can accumulate on the bed downstream because it is no longer flushed away by high flows. This fine sediment can infiltrate spawning gravels and reduce incubation success (for sediments finer than about 1 mm) or affect the ability of fry to emerge from the gravel (for sediments of 1 to 10 mm in size) (Kondolf 2000). This effect has been documented in many rivers, including in the Trinity River below Lewiston Dam, a notable example for

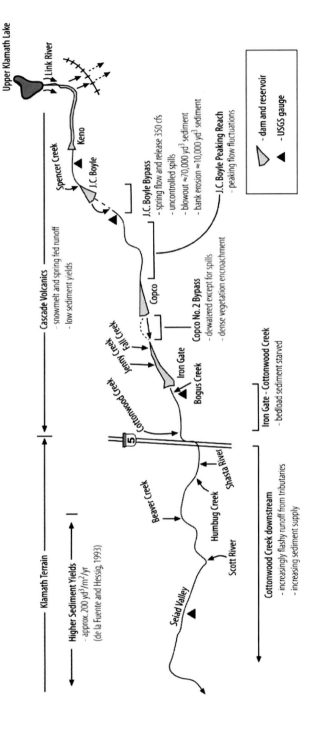

FIGURE 2-3 Conceptual model of the sediment transport and channel geomorphology in the Klamath River in the reaches affected by the PacifiCorp hydroelectric project dams.
SOURCE: PacifiCorp 2004. Reprinted with permission; copyright 2004, PacifiCorp.

the present study because it is one of the best-documented examples of this impact and because it is an important tributary to the Klamath (Milhous 1982). Reservoirs whose capacity is relatively small in relation to river flow typically allow high flows to pass while still trapping gravels supplied from upstream. Downstream of such reservoirs, the bed may progressively coarsen as the smaller gravels are transported downstream without being replaced (as they were before the dam was constructed) by gravels supplied from upstream. As a result, the bed may become dominated by larger gravels and cobbles that are unsuitable for use by spawning fish (Kondolf and Matthews 1993).

Reservoirs also can cause downstream changes in the distribution of riparian vegetation resulting from changes in hydrology and the availability of sediments. Reduced flood flows can result in less active bed scour, erosion, deposition, and channel migration, thereby resulting in smaller areas of fresh sediment surfaces available for colonization by seedlings of woody riparian species, but also in less frequent scour and removal of seedlings from the active channel. Thus, riparian vegetation can invade formerly scoured areas of the channel bed, but over time the riparian community may tend toward older individuals and later successional-stage species with less diversity of species and structure (Johnson 1992). Even if reservoirs do not substantially affect the high flows that erode and deposit sediment, they may affect the shape of the hydrograph during the seasons that riparian seedlings would normally become established, resulting in changes in the extent of riparian vegetation. Moreover, changes in water quality (from upstream land uses or transformations within reservoirs) can affect the growth of riparian vegetation through supply of nutrients for plant growth. Riparian vegetation is important as a resource in its own right, especially as it can provide important habitat to terrestrial and aquatic species. It also can affect geomorphic channel processes by increasing hydraulic roughness, by inducing deposition on bars and along channel margins, and by changing the direction of flow.

Changes Caused by Main-Stem Dams

The probable effect of the Pacificorp hydroelectric project reservoirs on various reaches of river is summarized in Figure 2-3. To understand the effects of these dams, it helps to recognize that they are small compared with the river's annual runoff. Upper Klamath Lake is large, but it is mostly natural, and its outflow is controlled not for hydroelectric production but for irrigation by the U.S. Bureau of Reclamation (USBR) in its operation of the Klamath Project. Keno Reservoir is an artificial structure at the site of a natural "reef" or bedrock sill that historically acted as a hydraulic control for Lake Ewauna, and the impoundment above J.C. Boyle Dam is essentially a forebay. By far the largest reservoirs are Iron Gate and Copco,[1] but even they impound only 4% and 5% of annual runoff, respectively. These relatively small impoundment ratios probably would not affect high flows substantially, except in bypassed reaches (reaches in which flows are reduced by diversion through penstocks for hydroelectric generation, such as the J.C. Boyle bypass reach and the Copco No. 2 bypass reach). As a result, the effect of the hydroelectric dams would be more likely to cause bed coarsening than accumulation of fine sediment (PacifiCorp 2004).

As discussed further in Chapter 3, there were important changes in how floods were routed between Upper and Lower Klamath lakes a century ago. Construction of the railroad embankment (and USBR control gates) blocked flood overflow into Lower Klamath Lake, as had occurred formerly. Current USBR irrigation facilities are managed so that in a flood situation Upper Klamath Lake water can be moved to the Lost River system. Water also can be evacuated from Keno Reservoir to the Klamath Irrigation Project via the Ady canal. Although data are not available, it is reasonable to conclude that elimination of flood overflow into Lower Klamath Lake would have increased the magnitude of flood flows in the Klamath River below Keno over that of conditions prevailing before the late nineteenth century. The increase in the magnitude of floods in the main stem also would tend to produce coarsening of the bed.

In bypassed reaches, the net effects of the dams would depend on the degree to which floods of various magnitudes have been reduced and on the base-flow conditions in the reach. For example, the relatively low (10 cfs) base flow maintained in the Copco No. 2 bypass reach, combined with changes to relatively short return-interval flood flows, has resulted in significant riparian vegetation encroachment. However, any such effects in the J.C. Boyle bypass reach, where the base flow is higher (100 to 300 cfs) and flood flow conditions are similar, are much more subtle.

Changes Caused by Tributary Dams

Dams on tributaries also are important and should be analyzed more comprehensively in the search for solutions to threats to fish populations. Irrigation storage dams on the Shasta River system result in large increases in water temperature and nutrient loads and are being studied by the University of California, Davis. The largest Klamath River tributary is the Trinity River, which has been regulated since the early 1960s by Trinity and Lewiston dams for power production and transfer of water to the Sacramento River system for irrigation, as part of the Central Valley Project. Approximately 51% of the Trinity's flow on average is transferred to the Sacramento River,

---

[1]This refers to Copco 1; Copco 2 is a run-of-the-river dam without substantial storage.

based on five water-year types as described in USFWS/HVT (1999) and currently managed by the Trinity River Management Council. By the 1970s, an abrupt decline in wild anadromous fish populations led to the first studies and restoration efforts. Continued studies and management manipulations have led to increased releases and a hydrograph more similar to a natural hydrograph in attributes important for fish life histories and to other efforts to restore dynamic channel processes, such as addition of gravel in sediment-starved reaches below dams (USFWS/HVT 1999).

Changes in Fish Populations

Steelhead run size before the 1900s is thought to have been up to several million fish per year. In 1960, run size was estimated at 400,000 fish, and the numbers continued to decline through the 1970s, 1980s, and 1990s. Returns of hatchery fish to Iron Gate Dam reflect an index of abundance and survival. From 1963 to1990, the average number returning was 1935 fish per year. From 1991 to1995, the average was 166 fish per year and in 1996, only 11 steelhead returned to the hatchery (as summarized in Hardy et al. 2006a). The Klamath Mountain Province steelhead populations do not appear to be self-sustaining and if trends continue, endangerment is possible. Steelhead have not been listed under the Endangered Species Act.

Coho salmon annual spawning escapement to the Klamath River system was estimated to be 15,400 to 20,000 fish in 1983 (Leidy and Leidy 1984). That estimate is less than 6% of their estimated abundance in the 1940s, and a 70% decline has been observed since the 1960s (CDFG 1994, as cited by Weitkamp et al. 1995). Coho salmon returns to Iron Gate Hatchery ranged from 0 in 1964 to 2,893 fish in 1987, and they are highly variable.

From 1915 to 1928, total annual harvest and escapement of Chinook salmon in the Klamath River was between 300,000 and 400,000 (Rankel 1982). In 1972, numbers were estimated to be 148,500 (Coots 1973). From 1978 to 1995, the average annual fall escapement, including hatchery fish, was 58,820 with a low of 18,133 (CDFG 1995). Spring Chinook salmon runs appear to be only remnants of the historical numbers. The numbers of Chinook spawning salmon, both wild and hatchery produced, from 1978 through 2006 are shown in Figure 2-4.

The Iron Gate Hatchery was established in 1963 at river mile 190 to mitigate the effects of the dams on anadromous species. Production goals for the hatchery include 4,920,000 Chinook salmon smolts, 1,080,000 Chinook salmon yearlings, 75,000 coho salmon yearlings, and 200,000 steelhead yearlings (Richey 2006).

The decline in numbers of anadromous fishes in the basin is commonly attributed to a list of anthropogenic factors, such as flow alterations due to

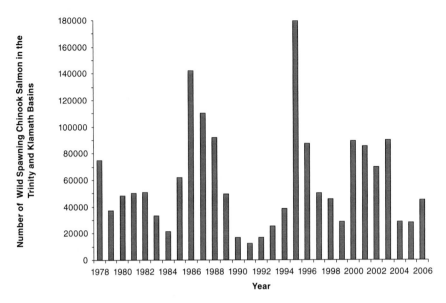

FIGURE 2-4 Spawning Chinook salmon, 1978-2006.
SOURCE: USFWS 2006, unpublished data.

irrigation and hydropower; temperature alterations as a result of riparian shading decreases and flow alterations; and land-use practices that alter in-stream habitat and contribute to sedimentation, including logging, mining, and agriculture. Other landscape factors that contribute to reproductive success and survival include fires, climatic changes, floods, droughts, and El Niño (Table 2-7). Other biological factors include reduced genetic integrity from hatchery production, predation, disease, and competition or preda-tion by introduced species. No single factor is known to be responsible for the decline in populations. It probably is the combination of the impacts and the timing of the impacts that can influence the productivity of these anadromous species. Thus, the pathway to reversing the trend of declining numbers is to determine the magnitude of the influences of the various fac-tors and to set priories for restoration efforts accordingly.

SUMMARY

The Klamath River basin is a complex hydrologic, geomorphic, and biological system with two sharply different sub-regions. The upper Klam-ath basin, the portion of the system upstream from Iron Gate Dam, includes extensive source areas for surface runoff along with irrigated agricultural

TABLE 2-7 Overview of the Habitat Factors Considered Important in the Decline in Anadromous Fish Populations and Their Potential Impacts

Factor	Impact to habitat	Impact to fish
Dam construction	(a) Limits access to upstream habitat (b) Alters habitat upstream and downstream by the creation of reservoirs upstream and sediment deficit downstream leading to a channelized or armored zone	(a) Reduction in available spawning and rearing habitat thus a reduction in potential juveniles produced (b) Creation of habitat suitable for nonnative species upstream and a decline in suitable instream habitat downstream through zone of influence
Flow alteration in amount, timing, duration, magnitude, and frequency (both tributary and main stem; resulting from water and land management)	Reduced channel-shaping flows; reduced maintenance flows for instream habitat; increased river temperatures; altered annual seasonal pattern; reduced base flows; can observe cumulative impacts to basin when multiple tributaries are affected	Reduced available habitat for spawning due to decreased water depths or changes in substrate conditions; stranding of juvenile fishes when flow changes are rapid; dewatering of redds
Timber harvest without proper consideration for riparian or watershed dynamics	Increased delivery of sediment to the channel, especially tributaries	Effects on spawning and changes in channel habitat
Placer, gravel, suction mining	Removes gravel from channel, resulting in changed geomorphology and instream habitat	Reductions in spawning gravel and homogenization of habitat types, leading to a reduction in carrying capacity for juvenile fishes and/or a decline in invertebrate production
Fires	Increased sediment delivered to streams	Effects on spawning and changes in channel habitat
Predation by nonnative species and mammals	No habitat impacts	Excessive predation may result in higher mortality rates than are sustainable by the population
Land use practices such as agriculture	Changes in riparian zone reduce shading especially in tributaries resulting in increased temperatures; increased delivery of sediment; increased delivery of nutrients or chemicals to stream channels	Increased algal blooms can result in dissolved-oxygen deficits; decreased usable habitat due to increased temperatures; sedimentation decreases spawning habitat, macroinvertebrate production, and changes in juvenile habitat

TABLE 2-7 Continued

Factor	Impact to habitat	Impact to fish
Commercial exploitation	No impact to habitat	Repeated reduction in spawning stock below levels needed for sustained support of the population results in reduction in production of juveniles
Climate change	By changing flow volumes and water temperatures, climate change would affect the availability, amount, and quality of riverine habitat available	Reduction in habitat availability and amount would reduce fish productivity; increases in late-summer temperatures could increase the frequency and magnitude of mortality events

areas, partly served by the Klamath Project of the USBR. Rivers are alluvial streams flowing across valley floors blanketed with alluvium that includes fine materials as well as gravels. A prominent feature of the upper basin is Upper Klamath Lake, a natural temporary holding area for runoff before it exits downstream through the Link River. The Link River Dam serves as a partial valve on the lake. The lower Klamath River passes through a mountainous area that has little agriculture but that is graced with steep forested slopes. Unlike the upper basin, the river is confined between bedrock walls and has a relatively steep gradient with a gravel bed.

These two physical provinces of the Klamath River basin host differing fish populations. The upper basin includes the lake and its several species of suckers (including the federally endangered shortnose and Lost River suckers), and tributary streams served as spawning areas for steelhead, coho salmon, and Chinook salmon, but dams in the system have cut off these streams from direct ocean access for fishes. The lower Klamath River and its tributaries served as habitat for several species of trout, salmon, and sturgeon among many other species. Dams now limit their access to the system to the area downstream from John C. Boyle Dam except for those fish able to ascend the fish ladder on the dam.

A variety of human factors, including changes to the basin hydrology, construction of dams, introduction of contaminants, logging of riparian forests, and fishing have contributed to the decline in the populations of many native fishes. Natural environmental changes, including those related to drought and flood frequencies and water temperature, are likely also to have affected populations of fishes. Where once the runs of anadromous species numbered in the millions of fish, present observations reveal less than 10% of historical numbers. Coho salmon are federally listed as a threatened species.

In an effort to understand these conditions and to improve them for the enhancement of fish populations, the USBR commissioned two models that were to provide insight to flows in the Klamath River and their effects on fish habitat: a reconstruction of what "natural" flows might have been like without dams in place (the Natural Flow Study), and a model to predict the distribution of fish habitats (the Instream Flow Study). These models operate within the general physical matrix described in this chapter, and they deal with the fish populations defined here. The remainder of this report explores the products of these research models.

3

Formulating and Applying Models in Ecosystem Management

"All models are wrong, but some are useful." G.E.P. Box.

INTRODUCTION

The documents the committee was charged to review are largely based on models. Models come in many different shapes and sizes, and the ways they are and can be used to inform management decisions vary enormously as well. Therefore, this chapter provides an overview of formulating and applying models in ecosystem management. It begins with a general overview and then progressively focuses on models used in aquatic, and especially riverine, ecosystems. Because there often is controversy over the appropriate role and use of models in decision making, the chapter concludes with discussions of the essential role of model testing and evaluation and use of institutional models for integrating knowledge and management. More detailed discussion of the models that underlie the National Flow Study (NFS) are in Chapter 4 and of the ones that underlie the Instream Flow Study (IFS) are in Chapter 5; in addition, a detailed discussion of models for use in regulatory decision making is in a recent National Research Council (NRC) report (NRC 2007a), much of which is relevant to the present case.

Modeling is the fallible art of trying to represent enough of the complexity and processes of real systems to solve a particular problem. Scientifically, such representations provide an ability to assemble more complex understandings of complex real systems than would be possible without such aids. They can be used to develop hypotheses that integrate many aspects of complex phenomena. Moreover, application of models can allow better predictions of the outcomes of proposed actions. This use of models sometimes allows more rapid, less costly, and less risky solutions to practical problems to be developed virtually than direct experimentation allows with the real system, especially for the systems discussed below.

Models have become indispensable for managing complex systems ranging from transportation systems (including most airline scheduling) to large building structures, as well as routine wholesaling, retailing, and commercial systems by engineers, business managers, and economists. In the physical and environmental sciences, conceptual and quantitative models have been central to the development of new theories and practices, especially in attempts to understand cause-and-effect relations in managed river systems, as well as in predictions of how natural systems will behave.

Historically, the scientific use of quantitative models began as early as the 1600s in Galileo's time, and engineering applications became established in France before the the the beginning of the French Revolution in 1789. Modeling now is the accepted approach for improving the efficiency and effectiveness of efforts to understand and manage complex problems. To improve the likelihood that modeling will deliver on such promises, model development and use commonly follows a fairly standardized process, described in this chapter. Scientific progress results when the hypothetical understanding of the system represented by the model diverges from field observations, leading to improvements in the model, field data, understanding of the modeled system, and the model's predictive powers.

Conceptual Versus Simulation Models

The science of river restoration is still in its infancy. In most river or wetland systems, there is only a partial understanding of the relation between flows, people, and ecosystems (Castleberry et al. 1996), and therefore science cannot yet provide certain predictions about the consequences of policy and management decisions. For this reason, the concept of "learning by doing" has become an accepted part of management activities in many river basins. A key part of the learning-by-doing process is the development of models that can be tested and refined through monitoring and research programs. Examples where modeling plays a prominent role in ecosystem restoration include the CALFED Bay-Delta Ecological Restoration Program (Healey et al. 2007), the Glen Canyon Adaptive Management Program (Walters et al. 2000), the Comprehensive Everglades Restoration Plan (Ogden et al. 2005), and the Trinity River Restoration Program (USFWS/HVT 1999, Schleusner 2006).

For the purposes of this discussion, the committee distinguishes between *conceptual* models and *simulation* models. Conceptual models serve to organize knowledge and information in the most general way, whereas simulation models attempt to describe system behavior quantitatively, using a series of deterministic or stochastic relations that link processes together to explore outcomes of different scenarios. The two types of models are often developed in tandem, conceptual models being used to lay the groundwork for restoration and for developing simulation models and simulation models being used subsequently to examine potential responses of system components. An example of this approach is given in the strategic plan for the CALFED Bay-Delta Ecological Restoration Program (CALFED 2000):

> Conceptual models are simple depictions of how different parts of the ecosystem are believed to work and how they might respond to restoration actions. These models are explicit representations of scientists' or resource managers' tacit understandings and beliefs. Conceptual models are then used to develop restoration actions that have a high likelihood of achieving an objective while providing information to increase understanding of ecosystem function and, in some instances, to resolve conflicts among alternative hypotheses about the ecosystem. The process of adaptive management can be enhanced when conceptual models are developed into simple computer simulations that can be used to explore the consequences of alternative options for restoration.

The description implies that conceptual models need not be particularly elaborate or precise; their primary purpose is to provide a framework for testing hypotheses and/or to coordinate research or restoration activities within complex systems. Figure 3-1 shows an example of a conceptual model illustrating the landscape of the Central Valley of California. The components of the landscape are represented by a series of boxes, with links between the boxes indicated by arrows. The arrows imply directional pathways, suggesting that processes or actions in one component of the model have the potential to generate a response in another component of the model. Scientists, resource managers, and landowners can (and often do) argue about the importance of the links, but recognizing their existence arguably is the most important step in developing ecosystem restoration strategies. Simulation models go a step further in representing landscape processes and interactions through computer algorithms and subroutines that quantitatively describe how the physical, biological and engineered components of the system interact in response to changes in state variables, such as water flow, sediment transport, and nutrient loading. Simulation models often fail to replicate landscape, riparian, or aquatic processes completely, but they are nonetheless useful because they permit exploration of general trends or serve to demonstrate the connections among a variety of measurable variables describing the physical and biological systems. Ecological modeling often is difficult to operationalize, but if substantial supporting data are available, such models can successfully replicate basic characteristics as water temperature, cross-sectional profiles, and flow velocity. Often, the most difficult task is to establish direct quantitative

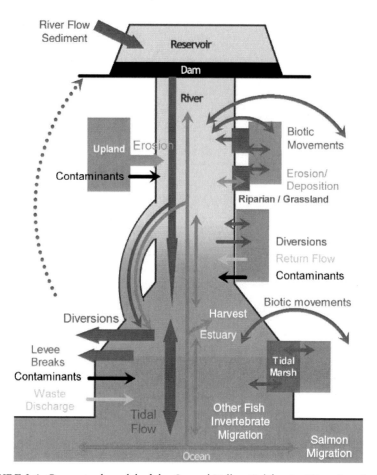

FIGURE 3-1 Conceptual model of the Central Valley, California. Diversions include diversions for agriculture.
SOURCE: Kimmerer et al. 2005. Reprinted with permission from the authors; copyright 2005, *San Francisco Estuary and Watershed Science.*

connections between the model that describes the hydrologic and hydraulic properties of the river and the ecological requirements of fishes or other aquatic organisms.

Examples of connections between flow and ecological models include applications of model strategies to the Colorado River downstream from Glen Canyon Dam, Arizona. Figure 3-2 shows a flow chart of the Grand Canyon Ecosystem Model, which was developed as part of the Glen Canyon Adaptive Management Program to examine how changes in the operation of Glen Canyon Dam will affect physical, biological, and cultural resources of the Colorado River (Walters et al. 2000). This model is an

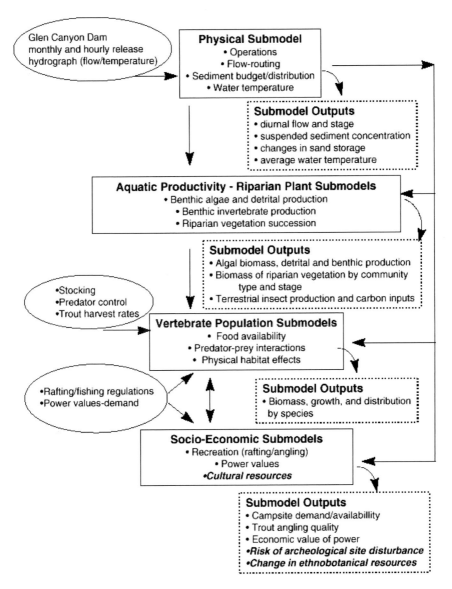

FIGURE 3-2 The Grand Canyon Ecosystem Model.
SOURCE: Walters et al. 2000. Reprinted with permission; copyright 2000, *Ecology and Society.*

executable computer program (in Visual Basic) consisting of separate sub-models that simulate the response of system components (boxes) to changes in reservoir operations, recreation activities and power demand (ovals). The model was developed through a year-long process that involved repeated meetings with scientists, managers, agency officials, tribal representatives, and advocacy groups, who collectively defined the scope of the problem and key modeling issues. The meetings served to not only parameterize the model but also to provide a mechanism for the various interest groups to express their opinions and reach consensus on the model framework and application. Subsequently, the Glen Canyon Monitoring and Research Center was established to collect and maintain the data and information necessary to test the model and further refine its application to managing the Grand Canyon ecosystem.

In this report, the committee is concerned with four specific kinds of models. The first three provide important driving variables for a model of the freshwater dynamics of salmonid fish populations:

• A *hydrologic* model that attempts to reconstruct pre-diversion natural flows of the Klamath River, drawing on historical hydrological data, measured physical relationships, and water balance calculations.
• A *water temperature* model used to simulate average water temperatures in a linear fashion down the main-stem Klamath River.
• A *habitat-suitability* model that predicts physical aspects of habitat for aquatic species as a function of stream flow.

The fourth model is a *fish-population* model that simulates salmon spawning, egg incubation, fry and juvenile growth, movement, survival and emigration to the ocean.

The first three models were formulated somewhat independently and they address very different questions and concerns. The fish-population model attempts to integrate these models by providing model linkages. This integration is limited by the different time steps among the physical models.

TYPES OF MODELS AND MODELING AREAS

Hydrologic Modeling

Often, hydrologic data needed for planning and design of water resources systems are either inadequate or unavailable at locations where the projects are built and operated. In such situations, engineers and scientists must rely on models to provide information for decision making. Hydrologic simulation models entail the mathematical descriptions of the components and the response of the hydrologic system (watershed or basin) to a series of events during the desired time. The resulting simulation models describe the various phases of the hydrologic cycle by using the laws of conservation of mass, energy, and momentum. The development and use of deterministic watershed-simulation models require a thorough understanding of the functions of the various components of the hydrologic cycle, as well as an adequate characterization of the spatial and temporal heterogeneities in the processes and the landscape.

Generally, a hydrologic simulation model consists of several sub-models that represent different components of the land phase of the hydrologic cycle. These sub-models usually consist of a set of nested relations that account for inputs, outputs, internal fluxes, and storages of water (Fleming 1974). The relevant hydrologic processes of the land phase of the hydrologic cycle vary substantially from one region to another. In high-elevation basins in the Pacific Northwest, for example, about half of the annual precipitation falls as snow (Serreze et al. 1999); thus, it is important to monitor seasonal changes in the extent and thickness of snow cover. Similarly, in arid and semi-arid regions, the water that potentially goes into the atmosphere via evaporation and transpiration is typically much greater than the water that falls on the surface as precipitation. Seasonal fluxes of water as a result of these processes, as well as groundwater flow and agricultural withdrawals, are particularly important in areas such as the upper Klamath River basin.

Since the development of the Stanford Watershed Model during the 1960s by Crawford and Linsley (1962, 1966), many hydrologic simulation models have been developed (Singh 1995, Wagener et al. 2004, Singh and Frevert 2006). Hydrologic simulation models for watersheds can be classified in many ways, and Singh (1995) provides a scheme based on process description, scale, and technique of solution. Most classifications use various adjectives to characterize the models according to the modeling properties. Commonly used adjectives that are relevant for hydrologic modeling in the Klamath basin are given in Table 3-1.

The structure of a hydrologic simulation model for a watershed or river basin can be simple or complex, depending on how close the degree of conceptualization of the hydrologic components is to the physical reality. Several comparative studies of different hydrologic models can be found in the literature. In 1975, the World Meteorological Organization (WMO) compared several groups of models, including explicit moisture-accounting models, such as the National Weather Service River Forecast System (NOAA 1972); implicit moisture-accounting models (also called tank models); and index models, such as the Antecedent Precipitation Index (API) model (WMO 1975). This study concluded that all models perform equally well on humid basins; that explicit moisture-accounting models are superior in semi-arid and arid areas; and that for poor-quality data, simpler models appear to give "better" results, primarily because the complex models have difficulties in accounting for changes in the soil-moisture balance.

The decision regarding the best approach for hydrologic modeling de-

TABLE 3-1 Adjectives Used to Classify Hydrologic Models

Adjective	Description
Black box	Process descriptions are based on appropriate mathematical functions fitted to data without any regard to the actual physics of the process
Conceptual	Process descriptions are based on various conceptualizations of the components of the hydrologic cycle
Continuous	Process is simulated for a long period, which usually includes many storm events. Moisture accounting is used to simulate the state of the process at the beginning of each event
Deterministic	Processes can be predicted with certainty without any random component
Distributed	Process descriptions account for variation of watershed characteristics from point to point
Event	Given the initial state, the process is simulated only for a single storm event of interest
Lumped	Process description ignores the spatial variation of watershed characteristics
Stochastic	Process is governed by random phenomena and the theory of stochastic process is used for its description

pends on many factors, including the availability of a modeling code for the problem at hand, data, resources, and time. In the Klamath basin, the contribution of groundwater to the total annual runoff may be a critical factor, especially as it influences stream flow recession that carries over from 1 year to the next. In addition, agricultural pumping within the basin might affect the shallow groundwater aquifers, which in turn might affect baseflow. The consideration of the role of groundwater will determine whether a model needs an explicit groundwater component (for example, the MODFLOW model from the U.S. Geological Survey [USGS]).

Based on the general requirements as outlined in both the Natural Flow Study (USBR 2005) and the Instream Flow Study (Hardy et al. 2006a), several candidate models could be considered for the hydrologic modeling in the Klamath basin. Table 3-2 presents these models along with some of the key characteristics that might help to choose among them. The selected model code should incorporate the processes needed to model the physical system accurately and to provide the information needed to satisfy modeling requirements. Typically, the models provide flowcharts for determining whether the features necessary for the particular watershed are included. Figures 3-3 and 3-4 provide examples of flowcharts and conceptual diagrams for PRMS and MIKE SHE models, respectively.[1]

---

[1]PRMS is a precipitation-runoff modeling system available from the USGS at http://water. usgs.gov/software/prms.html; MIKE SHE is an integrated hydrologic model developed by the Danish Hydraulic Institute available at http://www.crwr.utexas.edu/gis/gishyd98/dhi/mikeshe/ Mshemain.htm.

TABLE 3-2 Models for Coupling with the USGS Three-Dimensional (3-D) Groundwater Model MODFLOW

Model Code	Type	Surface Water	Groundwater	Time Scale	Spatial Scale	Reference	Used in Klamath Before?
GSSHA	Distributed, event, and continuous	2-D overland flow, 1-D stream flow	2-D, fully coupled	Variable, typically <1 day	Gridded	Downer et al. 2006	No
HEC HMS/RAS	Lumped, event, and continuous	Hydrologic methods for overland flow, 1-D stream flow	Hydrologic methods	Variable, typically <1 day	Lumped sub-basins, cross sections in canals	USACE 2007	No
HSPF	Lumped, continuous	Hydrologic methods	Hydrologic methods	≤1 day	Lumped sub-basins	Donigian et al. 1995	No
HYDRO-SPHERE	Distributed, event, and continuous	2-D overland flow, 1-D channel flow	Up to full 3-D		Gridded	Therrien et al. 2004	Yes?
MIKE SHE/ MIKE 11	Distributed, event, and continuous	2-D overland, 1-D channel	Up to full 3-D	≤1 day	Gridded	Graham and Butts 2006	Yes?
MODHMS	Distributed, event, and continuous	2-D overland, 1-D channel flow	Up to full 3-D MODFLOW	≤1 day	Gridded	HydroGeoLogic Inc. 1997	No
PRMS	Lumped, continuous	Hydrologic methods	Hydrologic methods	1 day	Lumped sub-basins	Leavesley et al. 2006	Yes
WASH123D	Distributed, event, and continuous	2-D overland flow, 1-D canal flow	Full 3-D	≤1 day	Gridded	Yeh et al. 2006	No

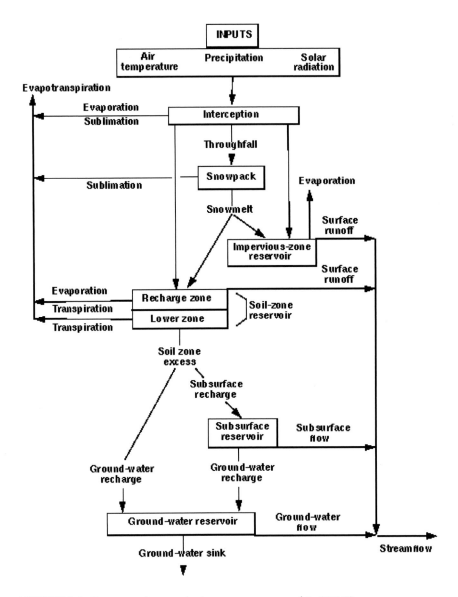

FIGURE 3-3 Conceptual watershed system represented in PRMS.
SOURCE: Leavesley et al. 2006. Reprinted with permission; copyright 2006, Taylor and Francis Group.

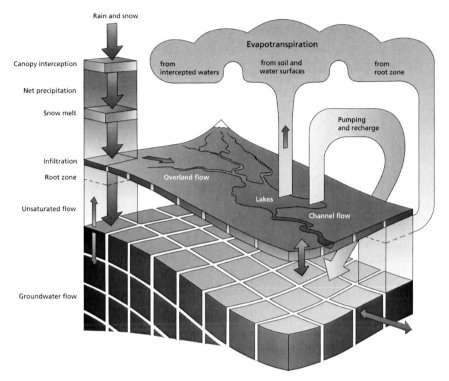

FIGURE 3-4 Schematic representation of a watershed in MIKE-SHE model.
SOURCE: DHI 2006. Reprinted with permission; copyright 2006, DHI Group.

The calibration of a hydrologic model is an extremely important step. Model results are only as good as the model itself, its input data, and its selected parameters. Models typically have two types of parameters (Sorooshian and Gupta 1995): physical parameters and "process" parameters. Physical parameters represent measurable properties of the watershed, such as area and slope. Process parameters are not directly measurable and depend on the particular scales (temporal and spatial) used in the model. Consequently, such parameters need to be determined through a process of "model calibration."

Two common model calibration criteria may be identified. First, the calibrated model must be able to reproduce the recorded historical data satisfactorily. Second, the parameter values of the calibrated model must be consistent with the watershed characteristics. This consistency can be verified effectively if the model parameters are directly related to measured physical parameters in the watershed. Usually that is the case with highly complex models that attempt to mimic the physical processes. During the early days of rainfall-runoff modeling, calibration was done "manually," but increased computing power now allows "automatic" calibration for some models, speeding up the calibration process and increasing objectivity and confidence in model predictions (Sorooshian and Gupta 1995).

Two other important steps associated with model development are model validation and model verification. Although many definitions, sometimes conflicting, have been suggested in the literature, both forms of model testing are intended to provide the assurance that the model code has been executed correctly and performs adequately for other data sets not used for calibration. The model calibration may also include another important step, calibration sensitivity analysis, to assess which parameters are important with respect to model performance and to estimate how well the parameters are identified (Wagener et al. 2004).

The use of the standardized modeling protocol for the development of hydrologic simulation models in the Klamath basin will provide many benefits. First, the calibration and verification steps provide confidence that the model can represent the physical system accurately. The use of a method that does not include these steps (for example, naturalization of stream flows using historical data) might result in unreliable predictions with significant uncertainties. Second, the standardized modeling protocol allows the analysts to use the models as predictive tools for scenario analysis. Two such scenario analyses are estimation of stream flows under natural (pre-development) conditions and projection of managed system flows under future developed conditions. If an uncertainty analysis of the model's structure, parameters, and inputs accompanies such predictions, then the modelers can provide guidance on the uncertainty of the results for decision making.

Temperature Modeling

The Instream Flow Study (USBR 2005) described the use of two different hydrologic and water-quality models. The first was the Systems Impact Assessment Model (SIAM) (Bartholow et al. 2003), which incorporates a water-quantity component (MODSIM) and a water-quality component (HEC-5Q) that are used to simulate water temperatures. A more complete description of the HEC-5Q model adapted for the Klamath River is given by Hanna and Campbell (2000).

The second modeling approach used simulations from PacifiCorp (2004) developed as part of the effort to relicense the Klamath Project and simulations from the North Coast Regional Water Quality Control Board (NCRQWB 2006). These efforts incorporated hydrodynamics (RMA-2) and water-quality (RMA-11) models, the latter including water temperature. Wells et al. (2004) provide a technical review of these models. The simulations were used in the Instream Flow Study "for examination of differences in inflows under their assumed 'natural flow conditions' for calendar year 2000 and existing conditions for the 2000 and 2002 calendar years." Based on the water-quality simulations conducted during the PacifiCorp relicensing and the ongoing total maximum daily load (TMDL), the Instream Flow Study states that "we believe that dissolved oxygen and other water-quality parameters are of secondary importance to our efforts compared to that of temperature."

Habitat Modeling

Two specific types of riverine habitat models were used in the Instream Flow Study. These are commonly referred to as micro-habitat models, representing the hydraulic aspects in two dimensions; and macro-habitat models, representing the water-quality aspects (temperature in this study) in one dimension (Bovee et al. 1998).

Micro-Habitat Modeling

Providing suitable hydraulic habitat conditions is a necessary part of any instream flow prescription, but it is not sufficient by itself (Annear et al. 2004). The modeling of the hydraulic component of physical habitat also is referred to as habitat-suitability or habitat-selection modeling and may also be considered a subset of resource-selection models (Manly et al. 1993, 2002). The use of habitat-selection models for predicting effects of habitat alteration is premised on the general assumption that the population response of interest (often year-class strength or abundance) varies in proportion to the availability of highly selected habitat types considered over time and space; highly selected habitat types are habitats where the species has been observed at high densities. Areas where individuals are rarely observed sometimes are termed "unusable," and models focus on how the area of highly selected and unusable habitat varies among management alternatives.

In general, habitat-selection modeling includes the following steps (Manly et al. 2002, Scott et al. 2002):

1. Identify the target animals (the species and life stage to be modeled).
2. Select a spatial resolution (cell size) appropriate for the target organism. This resolution must reflect the distances over which the animals select and use habitat. For territorial fish, the resolution should not be

smaller than a typical territory size; for nonterritorial fish (for example, fish that forage over whole pools and into adjacent riffles), the spatial resolution must be at least as large as a unit of habitat.

3.   Identify habitat variables expected to be useful in predicting animal presence, absence, or density. The variables should be ones that strongly affect the fitness value of habitat (for example, affecting food intake, metabolic energy costs, and predation risk) or habitat usability or avoidance, and should be readily measured over the chosen spatial resolution.

4.   Collect field observations. Divide the habitat into cells of the chosen spatial resolution and observe the presence, absence, and density of animals in each cell; and evaluate the habitat variables for each cell. Habitat variables must be observed for habitat occupied and unoccupied by the target animals. The observations must provide an adequate sample size for the subsequent statistical modeling and must attempt to include ranges of habitat variable values as extensive as those to which the model will be applied.

5.   Choose an index of habitat selection (a measure of how strongly the animals select a particular type of habitat). The simplest approach is to use the density of animals (number observed per unit area) or simply the presence of animals as the habitat-selection index (Manly et al. 2002).

6.   Use the field observations to fit a statistical model of how the habitat selection index varies with the chosen habitat variables. Typically this activity means modeling animal density as a nonlinear, multivariate function of the habitat variables, including interactions among variables (Manly et al. 2002). Model fitting should consider several statistical concepts, such as parsimony (the inclusion of habitat variables in the model only if they significantly improve the model's ability to explain the data) and overfitting (the potential for a model to be skewed by a few data points at the extremes of variable ranges).

7.   Apply the statistical model to altered habitat scenarios to predict how the habitat-selection index varies among management alternatives (for example, at different flows).

8.   Test model output and evaluate uncertainty in model predictions. The model-fitting step should produce uncertainty statistics that can be used to evaluate confidence bounds on predictions.

Procedures have advanced substantially since the 1970s, when PHABSIM was a pioneer habitat-suitability model. PHABSIM applications have been controversial at times, including when the habitat model was applied as a static index for selecting a single "minimum instream flow" in the absence of supporting hydrologic time series and hydraulic models. The PHABSIM model is intended to provide input to the spatial and temporal analyses within the more comprehensive decision-making framework of the Instream Flow Incremental Methodology (IFIM) (Bovee et al. 1998).

As discussed in the Scale Issues section below, the spatial resolution (cell size) used in PHABSIM studies often is not selected for biological reasons, but for ease of hydraulic modeling. Consequently, the field observations used to generate fish habitat criteria might be at a very different spatial resolution from that of the hydraulic modeling. Instead of having to decide which habitat variables are important to include, if they fail to follow the logic of the comprehensive framework of IFIM (particularly the water-quality aspects), PHABSIM-based instream flow studies often assume a priori that a few variables (usually depth and velocity and, sometimes, substrate type) are the only important habitat variables.

Some PHABSIM studies attempt to develop habitat suitability criteria from observations of habitat occupied by fish without considering the availability of unoccupied habitat. A meaningful model using data only from occupied habitat cannot be made without knowing how many and what kinds of habitat were available but not occupied (Manly et al. 2002). Of equal importance is an understanding of the kind of habitat conditions that are avoided (never occupied). Instead of explicitly modeling the density of fish in each cell, PHABSIM produces a "weighted-usable-area" (WUA) output, which is a function of usable habitat for specific life stages of aquatic species plotted against flow. Although similar to a density model, WUA output functions have no clear meaning by themselves and cannot be measured in the field. "Usable" is not the same as "occupied." A usable area can be unoccupied, but an unusable area, by definition, cannot be occupied. Only during testing of model output do statistical analyses differentiate between presence or absence of fish and the "usable" versus "unusable" or "optimal" versus "marginal" habitats (see Chapter 5 for detailed discussion of this matter). These functions are more appropriate as input to simulations of usable micro-habitat area coupled with usable macro-habitat area and species life-history periodicity to produce "total usable habitat" simulations. Such simulations incorporate hydrologic time series, water-quality modeling, and stream-channel characteristics to simulate space-habitat conditions over time and to identify potential habitat-imposed limitations or "bottlenecks" (if any) to fish populations by alternative water-management proposals.

Conversely, the simple observation of flow and habitat functions does not provide a basis for inferring the response to changes in flow. For example, there is no scientific basis for assuming that a doubling of usable habitat area would automatically result in a doubling of the survival, growth, abundance, or biomass of a specific life stage. The statistical modeling methods are discussed by Guay et al. (2000) and Ahmadi-Nedushan et al. (2006). The parsimony and overfitting issues are rarely addressed explicitly. The suitability-criteria approach can facilitate evaluation of model uncertainty only when incorporated as part of time-series simulations of habitat dynamics through time and space. PHABSIM suitability criteria are virtually never accompanied by goodness-of-fit statistics. Rather, testing of model output focuses on the comparison of independently collected observations of fish distributions with modeled usable vs. unusable habitat simulations representing flow conditions present during fish observations (Thomas and Bovee 1993). Habitat-selection modeling has received widespread recent attention in general (for example, Garshelis 2000, Burgman et al. 2001) and specifically in reference to instream flow assessment (for example, Orth 1987, EPRI 2000, Railsback et al. 2003). The fundamental assumption that populations respond in proportion to the availability of highly selected habitat is not well supported (Railsback et al. 2003). Fish populations are limited in part by factors (especially food availability) other than physical habitat. Competition within and among species for habitat can cause habitat selection models to be misleading. Habitat created for small fish can be occupied instead by larger fish or fish of another species (reported in field studies by Loar et al. 1985).

Because habitat-selection models do not consider time, they are not suitable for predicting effects of habitat changes over time. They produce different results for different life stages and different species, but there is no consistent way to combine these results into a meaningful prediction of overall effects on a species or community. For example, if a change in flow were predicted to double the area of selected habitat for salmon fry but halve it for larger juveniles, the results would not by themselves allow prediction of the overall effect on salmon production, although they would identify areas for further modeling and research.

Macro-Habitat Modeling

Macro-habitat modeling deals with "the set of abiotic conditions such as hydrology, channel morphology, thermal regime, chemical properties, or other characteristics in a segment of river that define suitability for use by organisms" (Bovee et al. 1998). The hydrologic and temperature regimes are of particular focus by the Instream Flow Study. The hydrologic time series is a chronological distribution of stream flow at particular locations in the stream network of interest. Temperature models produce biologically relevant information, such as degree-day accumulations, at specified locations along a river. The models produce steady-state longitudinal profiles of the temperature downriver. By superimposing biological information related to suitable temperatures (such as acute effects, length of egg incubation, and growth rates), the stream length of usable macro-habitat as a function of discharge can be simulated. These effects are usually evaluated parallel to the micro-habitat and can be coupled with hydrologic time series to provide times series of total usable habitat.

Fish-Population Modeling

Fish-population modeling involves the use of the fish-population model SALMOD (Figure 3-5), which was used in the Instream Flow Study to evaluate the instream flow recommendations by comparing simulated Chinook-salmon smolt production from the Klamath River under conditions of imposed flow recommendations rather than existing conditions. SALMOD is a computer model that simulates the dynamics of the freshwater phase of either anadromous or resident salmonid fish species (mainly Chinook and coho salmon and steelhead and rainbow trout in the Klamath system). Spatial resolution is consistent with the mesohabitat distribution through the stream reaches of interest. Stream flow, water temperature, and mesohabitat type are the physical variables included in the model. The biological resolution uses a typical categorization of fish life history related to physical morphology, behavior, and reproductive potential. The model has been used to predict the population consequences of alternative flow and temperature regimes, to understand the relative magnitude of mortality in determining the timing and degree of habitat "bottlenecks," to design flow regimes to mitigate habitat bottlenecks, and to explore the effectiveness of stocking programs. The SALMOD model is conceptually illustrated in Figure 3-5, and its application to the Klamath River is described by Bartholow and Henriksen (2006).

Hydraulic Modeling and Instream Flow Studies

Depths and velocity fields in natural rivers are complex and irregular, often having significant cross-stream components (Dietrich and Smith 1983, Petit 1987, Whiting and Dietrich 1991, Larsen 1995, Whiting 1997). This complexity in the flow patterns of natural channels poses a challenge for methods of assessing instream flows that depend on hydraulic modeling. Consequently, the melding of hydraulic and biological aspects in habitat-suitability modeling has been the subject of continuing criticism (for example, Marthur et al. 1985; Shirvell 1986, 1994; Osborne et al. 1988; Gan and McMahon 1990; Elliott 1994; Castleberry et al. 1996; Ghanem et al. 1996; Heggenes 1996; Williams 1996; Lamouroux et al. 1998). One- and two-dimensional hydraulic models and their application to instream flow studies were reviewed by Kondolf et al. (2000) and are summarized below. Figure 3-6 provides an overview of how several of these model types are used together.

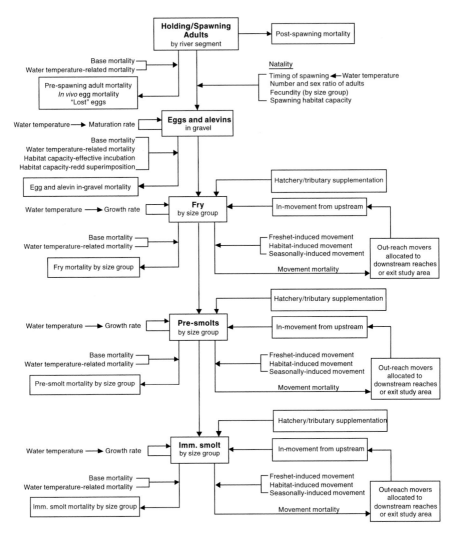

FIGURE 3-5 Conceptual illustration of the variety of factors important in controlling salmon production throughout SALMOD's biological year.
SOURCE: Williamson et al. 1993. Reprinted with permission; copyright 1993, *Regulated Rivers-Research & Management.*

One-Dimensional Models

Many one-dimensional (1-D) hydraulic models are step-backwater models, which apply the Manning's equation to calculate river stage for a given discharge, typically treating the river as a series of cross sections, for each of which the cross-sectionally averaged depth and velocity are com-

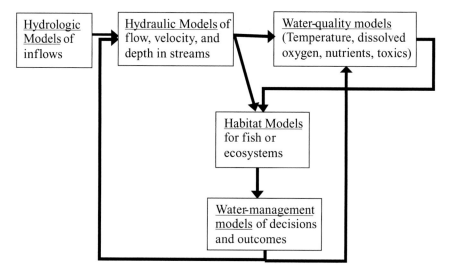

FIGURE 3-6 Interactions among models to represent an ecosystem and its management.

puted based on hydraulic principles, channel form, and a known relation between stage and discharge at the downstream hydraulic-control cross section. The best-known 1-D hydraulic models are HEC-2 and the newer HEC-RAS, which now are widely used for predicting flood levels. WSP, a similar 1-D gradually varied flow model, is an option for modeling stage in PHABSIM (Milhous et al. 1984).

One-dimensional models typically assume that the channel is straight, all flow being perpendicular to the cross section, and that flow is either "uniform" or "gradually varied." Uniform flow does not change in the downstream direction and, therefore, has a vertical velocity profile that reflects a balance between the acceleration of gravity and the resistance of the channel bed. Those conditions more commonly occur in canals but generally not in natural streams. Those conditions can be approximated by closely spaced cross sections. Gradually varied flow occurs where channel topography and roughness change only slowly along the channel, so that convective accelerations can be ignored.

These are important assumptions, and while reasonable approximations of river stage are routinely obtained with these models if they are used with adequate skill, they can provide only cross-sectionally averaged velocity. Moreover, gradually varied flow models are commonly used to predict flood stage during high flows. During high flows, variations in the bed topography may become less important, and hydrologists often speak of riffles being "drowned out" at high flow stages. Instream flow assessments, however, also are concerned with the lower-magnitude flows in which fish spend most of their time. These flows are too low to modify the streambed, so they occupy a channel geometry inherited from past higher flows. Downstream changes in channel geometry that are small relative to high flows may be large relative to low flows, as when low flow spills over a longitudinal bar, so that the assumption of gradually varied flow is violated (Osborne et al. 1988). As a result, a model that gives reasonable estimates of stage in a channel at high flows may fail to do so at low flows.

PHABSIM is concerned with the distribution of velocity and depth across the channel; therefore, the hydraulic sub-models in PHABSIM divide the cross section into vertical slices (cells) either centered on or between point measurements of velocity (much as is done in the USGS discharge measurements). The vertical cells are analyzed separately, using either a regression analysis of measurements of velocity in the cell at different stages or a back-calculation of Manning's *n* from a single velocity measurement (Milhous et al. 1989). In the latter approach, the cells are no longer tied to one another through hydrodynamic principles (Shirvell 1986, Ghanem et al. 1996). With the single-measurement approach, the Manning's roughness factor is used to calculate velocity and discharge for each cell at other discharges, but the individual cell discharges are adjusted to equal the modeled flow; therefore, the roughness factor is a calibrated weighting factor rather than a true roughness coefficient.

Two-Dimensional Models

Two-dimensional (2-D) models are increasingly being used for instream flow studies (for example, Leclerc et al.1995, Ghanem et al. 1996). Two-dimensional models require the simultaneous solution of a system of governing equations, typically including relationships (expressed as differential equations) for conservation of fluid mass, conservation of downstream fluid momentum, and conservation of cross-stream fluid momentum. To simplify these relationships, some approximations are assumed, yielding the "shallow water equations." These 2-D velocity models give only vertically integrated velocities but show the variation in cross-stream direction as well as in the downstream direction.

Two-D models retain the convective acceleration terms neglected by 1-D models but require more detailed descriptions of channel geometry, and the accuracy of the modeled results depends on the accuracy and spatial resolution of the measurements (Leclerc et al. 1995, Ghanem et al. 1996). For example, Leclerc et al. (1995) constructed a computer representation of the bed of a large stream by measuring the bed elevation with one measurement for every 50 to 400 m^2, and their results are necessarily generalized accordingly. With detailed specification of the channel bed topography and planform, more sophisticated modeling might not be necessary.

One-dimensional models are not all the same, and in some settings, 1-D models can be as accurate for simulating vertically integrated velocity fields as a 2-D approach. Dietrich (1987) modeled flow in Muddy Creek, Wyoming, for geomorphic purposes, using a 1-D approach that explicitly accounted for the effect of channel curvature, and predicted the distribution of velocity across the transects. Larsen (1995) applied the same 1-D approach and compared observed velocity patterns on two gravel-and-cobble-bedded meandering rivers. He showed that, with good bed topography as input, the 1-D model performed as well as more sophisticated models. Waddle et al. (2000) found similar ability to predict stream-velocity distribution when comparing 1-D and 2-D model simulations, provided that the bathymetry of the channel was adequately described (for example, transects were placed close together for 1-D model field-data collection). However, understanding the appropriateness and limitations of a model seems critical, and in a straight channel with irregular bed topography, such as that studied by Whiting and Dietrich (1991), a 2-D model that accounted for convective accelerations would be more appropriate.

Hydraulic models used as input for habitat time-series analyses must involve detailed testing of model-velocity output over a range of flows, including overbank, to establish credibility. Calibration to water-surface elevations alone is not sufficient.

Statistical Hydraulic Models

Following a suggestion by Dingman (1989), Lamouroux et al. (1995) developed an empirical model that predicts the statistical distribution of depth and velocity for reaches with intermediate and large roughness elements, for which they believe the conventional deterministic models are ineffective. The model predicts the distributions of the hydraulic variables over an entire reach based on inputs of discharge, mean width and depth, and roughness. Lamouroux et al. (1998) coupled this hydraulic model with multivariate habitat-use models to estimate the habitat value of a reach as a function of discharge. The need for validation is perhaps more obvious with such straightforwardly empirical models, which is a virtue.

Water-Quality Models

Ecosystems rely on more than just the hydrodynamic conditions of stream flows. Water quality, such as temperature, concentrations of dissolved oxygen, nutrients, toxic agents, and other substances, and their fluctuations are also important. Water-quality models integrate knowledge and understanding of water-quality processes (conservation of mass, erosion and deposition, reactions and reaction rates, and transport advection and dispersion) with results of hydrodynamic-model results that affect the movement and fate of water-quality constituents. Typically, hydraulic- or hydrodynamic-model results provide estimates of flow, on which water-quality movement and reaction processes depend. Water-quality models often are made to go along with, and sometimes reside inside of, hydraulic or hydrodynamic models.

Water-quality models typically rely on fundamental knowledge of the physics of water flow and its interactions with water-quality constituents, for example, temperature and chemical constituents, such as conservation of mass and energy, mixing, and forms of chemical and physical reactions. This fundamental knowledge is widespread and so can be coded in model software for wide application and is supplemented by local knowledge of chemical inflows, local reaction and mixing rates, and other constants. The boundary conditions and parameters are either estimated from field measurements, related laboratory studies, or drawn from studies of other locations with similar conditions.

Two water-quality models that were used in the Instream Flow Study are SIAM-MODSIM/HEC-5Q and RMA-2/11 (Hardy et al. 2006a).

Water-Management Models

The operation and performance of water resources are of great practical interest to the many economic and noneconomic users of these environmental services. Water-management models reflect the operational decisions and policies of the system, making operational changes in response to hydrologic, hydraulic, water-quality, or habitat conditions. These management decisions, in turn, most directly affect stream hydraulics (through the operation of dams and diversions) but, depending on the decisions, can also affect water-quality inputs (contaminant loads), temperature (from flow changes and riparian shading), and habitat (by land- and water-use decisions).

Water-management optimization models take a different approach, using automated mathematical procedures to suggest promising quantitative decisions that maximize quantitatively defined operating objectives for the system within constraints. Whether using simulation or optimization, water-management models integrate the various relevant aspects of the system to support investigations of particular (simulation) or potentially optimal (optimization) decisions and their evaluation in terms of quantitative assessments of performance.

For the Klamath basin, several water-management models have been applied. Linkages among the various models discussed, including CLASIM and MODSIM, can be used to provide input for water-management models as illustrated in Figure 3-6. For example, the linked modeling effort (SIAM) was constructed by the USGS to describe the Klamath River (Hanna and Campbell 2000).

Scale Issues in Ecological Modeling

Choosing appropriate scales is a crucial step in environmental and ecological modeling. Scales must be chosen to represent both space and time. Scale issues include dimensionality (which dimensions to represent), extent (the total area and time to be represented), and resolution (or "grain," the size of the slices of time and space).

For space, selecting dimensionality means determining whether a model is non-spatial (processes are assumed not to vary over space), one-dimensional (model variables and, possibly, processes can vary in one spatial dimension), or two- or three-dimensional (variables and processes vary in two or three dimensions). Adding dimensions to a model usually comes at a high cost in model complexity, so good models use no more dimensions than necessary. The choice of dimensionality depends on the processes and variables being modeled and the system; the key question is whether variability in any dimension is low enough that the model can ignore it and still solve the problem the model is intended for (Grimm and Railsback 2005). River temperature, for example, typically varies much more in the upstream-downstream dimension than it does in the vertical and cross-channel dimensions, so most river temperature models are one-dimensional. However, if the purpose of a temperature model includes predicting the extent and magnitude of thermal refuges from tributary inflows, then the model also must represent variation in the cross-channel dimension. If thermal refuges occur at the bottom of deep, stratified pools, then the vertical dimension must be represented. Fish-habitat models typically are two dimensional, ignoring vertical variation. Such habitat variables as velocity and distance to hiding cover do vary vertically, especially in large rivers, but this variation is typically smaller than the variation in horizontal dimensions and is not important enough to justify the substantial complexity of simulating it.

Choosing the spatial extent of a model means deciding the total area modeled. For river models, the necessary extent is usually well defined because the river has clear-cut edges. The extent in the upstream-downstream dimension is often clearly defined by dams, the river's mouth, and the distribution of the aquatic organisms of interest (including seasonal movement); however, for modeling water quality, it often is not clear how far upstream a model needs to start to represent dynamics at downstream sites. Selecting an appropriate spatial resolution is typically the least clear-cut scale issue, yet it is perhaps the most important (Grimm and Railsback 2005). For two-dimensional models such as those often used for instream flow assessment, spatial resolution refers to the size of the cells used to model habitat variables. Cell size is determined by the distance below which spatial variation can be ignored; the model assumes conditions are uniform within a cell. Clearly, one factor determining spatial resolution is the distances over which habitat variables change. A large river with large patches of relatively uniform habitat does not require small cells. However, for ecological models the organisms being modeled also are important determinants of an appropriate spatial resolution. Biological issues in selecting a spatial resolution include what distances the animals move for what reasons, over what periods. For strictly territorial animals, the territory size often makes a useful grid size.

Temporal-scale issues are analogous to spatial-scale issues in that one is deciding how to divide and represent time instead of space. The question of dimensionality is simply whether a model should represent time (a dynamic model) or ignore time (a static model). Models can be static if they represent processes and variables that do not vary over time, or if temporal variability can be ignored for the model's purposes. Habitat-suitability models (including the PHABSIM-like approaches) have been used in dynamic modeling (as in IFIM) and as static indexes, ignoring temporal variation. Static habitat indexes might not have been much of an issue during the early days of instream flow study, when decisions about water development typically called for a single minimal flow release applied year-round. Modern instream flow studies require dynamic habitat modeling for instream flow recommendations that include inter-annual variations (seasonal, monthly, daily, or even hourly) and inter-annual variations (extreme hydrographs that invade the floodplain, depositing sediments, providing biological refuges and rearing habitat, replenishing nutrients, recruiting large woody debris, and scouring sediments) (Annear et al. 2004). Consequently, the use of static habitat models no longer reflects the state of the science (NRC 2005b).

If a model is dynamic, then its temporal resolution (time step) must be chosen. The time step should be the length of time within which variation can be ignored, either because within-step variation is small or because capturing within-step variation is not important for the model's purpose. River temperature models are often dynamic, so they can capture how temperature varies with season and weather. If a temperature model's purpose is to predict mean daily temperature, then the model can operate at a daily time step, the input representing the daily average or the total solar radiation, air temperature, and so forth. If the model's purpose is to predict daily maximum temperature, hourly time steps often are used. Even shorter time steps would produce a more accurate prediction of maximum temperature but the additional accuracy often does not justify the additional effort to use shorter steps.

Among the most important yet difficult to understand scale issues is the potential for important errors and hidden assumptions to arise from mixing spatial or temporal scales. Many model parameters, inputs, and assumptions are valid only at certain scales, and using them at other scales or mixing model components across scales is a source of error common in environmental management modeling. Subtle changes in model meaning (or outright errors) can result from mixing spatial and temporal resolutions.

A well-known example of time-scale-dependent parameters is temperature criteria. For example, "What temperature is lethal to salmon?" is a meaningless question unless the duration of exposure is specified. A juvenile salmon might survive several hours at 26° C with no permanent effects but would be unlikely to survive for a day or more at that temperature. Hence, using temperature criteria requires examining the temporal scale for which they are appropriate—for example, a criterion useful for daily mean temperatures is not useful for hourly temperatures.

Similarly, questions such as "At what flow is habitat optimized?" or "At what flows do habitat bottlenecks occur?" also are meaningless without specification of the species' life-stage periodicity, extent of exposure, and place of occurrence. River-habitat conditions are dynamic, and habitat modeling of their temporal aspects is necessary to identify habitat bottlenecks that may be controlling fish populations by limiting one or more life stages. The spatial aspects of river habitat determine to a large extent where conditions for successful completion of fish life stages are likely to occur in particular river environments. Upstream and downstream passage and suitable spawning and rearing areas might be restricted to relatively small portions of specific river environments and might be substantially modified by the flow and temperature regimes experienced by the areas.

In summary, one of the key discoveries of riverine ecology in the past several decades is the importance of spatial scale and temporal scales. Ecological processes and parameters can differ strongly as scale varies (Levin 1992, Bissonette 1997), so models must use consistent, ecologically appropriate scales. Mixing spatial resolutions within a model (hydraulic model cells of several square meters vs. habitat criteria developed from data measured only a few centimeters from where fish are observed) is a fundamental flaw. The mismatch of scale among habitat suitability, hydrology, hydraulic, and temperature modeling will result in predictions of questionable utility. More narrowly focused scale issues on streams are discussed in Chapter 5.

THE MODELING PROCESS

Although many authors present somewhat different versions of the modeling process, most presentations support a process of developing and applying a model with elements and steps similar to those listed below (Gass 1983, Satkowski et al. 2000, EPA 2003):

1. Problem and purpose definition
2. Model development

3. Calibration and testing
4. Sensitivity analysis
5. Results interpretation and communication

These steps should lead to the selection of temporal and spatial scales of the information required, and also the level of accuracy of modeling output necessary for decision making. Figure 3-7 illustrates the process in more detail, starting with the development of conceptual and mathematical models; linkage to numerical computer algorithms; data collection, calibration, and verification; and, finally, reflections on model performance and suggestions for model refinements. A planned approach for the development and application of a model is greatly beneficial, as it is for other problem-solving activities (Polya 1957).

Problem Statement and Purpose

"A problem well posed is a problem half-solved."
(attributed to John Dewey)

Most models are developed for a particular scientific or problem-solving purpose. Defining a purpose for the model provides a focus and rationale for including factors and processes that are more likely to be important and for excluding others. Modeling the world in its entirety is never an option—the primary criterion for deciding what needs to be in a model is to include only the components necessary to solve the problem and little else. A well-defined applied model not only describes the real system by clarifying the problem that needs to be solved; it also can help in the development of an appropriate institutional framework for implementing solutions.

Defining the purpose of the model is the most important step in modeling. For applied modeling, a common error is modeling the wrong problem. Such errors can arise from adapting an older model—developed for a different purpose—to a new, narrower purpose, for example, when an expert seeks to apply this specialty to unsuitable circumstances, when political concentration is on one aspect of a problem when another dominates (as might be the case with Klamath studies concentrating on the main stem when tributary processes seem at least as important), and frequently when the nature of the problem changes in the years of model development (a lag between model development and use of model results). Sometimes it is impossible to tailor a model to a particular purpose without important simplifications; a simple but accurate model of a complex natural system would be a common example of a potentially impossible modeling purpose. Thus, models commonly are not perfectly suitable for the practical prob-

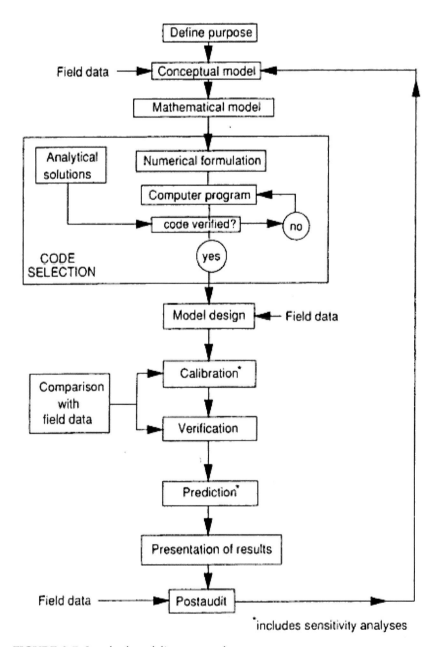

FIGURE 3-7 Standard modeling protocols.
SOURCE: Adapted from Anderson and Woessner 1992. Reprinted with permission; copyright 1992, Academic Press.

lems to which they are applied. Such problems often can be mitigated by changes in the model (perhaps recalibration) and careful interpretation and communication of model results, which are described later.

Model Development

Following a statement of the model's purpose, knowledge of processes and structures thought to be most relevant to the problem is assembled. This knowledge can take the form of empirical relationships, observed locally or in similar circumstances, and relationships derived from fundamental and well-proven principles. Conservation of mass, energy, and momentum are examples of fundamental principles from which relationships can be derived. Empirical relationships are inferred from field data by regression or by other types of fitting to equations. At this point, key variables or parameters that can be measured or evaluated in the modeling and testing process must be identified. Field data can be obtained locally or from locales deemed similar to develop empirical databases. Mathematical forms of these empirical and fundamental relationships are then organized into a coherent representation of the system for the purposes of the problem. Simplifications to some parts of the problem often are required to produce tractable forms. In modeling ecosystem responses to flow, it is essential to recognize spatial and temporal differences in scale because the relationships for individual organisms and hydrodynamic processes change markedly with scale.

This simplified representation of the problem sometimes must be simplified further to allow solution or approximate solution of the mathematical problem. Numerical methods, such as finite-element or finite-difference techniques, often are used to solve relatively complex mathematical representations. Frequent checks on the stability and accuracy of the numerical solution often are required.

At the end of this step, the model of the system is twice simplified from the original real problem, first to create a mathematical representation of the problem and then to create a solvable mathematical representation. Nevertheless, the result is commonly a far more complex and transparent representation of the problem than would be possible without mathematical aid, and a representation that allows integration of diverse types of scientific knowledge and understandings of the system.

Calibration

A further empirical phase of model development is model calibration. Calibration consists of adjusting some of the more empirical parameters in the mathematical model to fit data observed from the field. Sometimes parameters in component sub-models are adjusted against field data and sometimes parameters in several model components are adjusted together against field data. Sometimes calibration is based on data observed in field conditions elsewhere, if local data are unavailable. Having local field data is greatly preferred under problem-relevant conditions to calibrate empirical parameters. However, field data rarely are available to the extent desired within a time frame relevant for the problem.

The adjustment of parameters often is done by experts in modeling the type of system being modeled. Such adjustments sometimes are aided by automated algorithms, particularly when calibration parameters are numerous. Because often there are many possible sets of parameter values that "fit" field data, the background and understanding of the modeling experts have an important role in calibration. Usually, parameter calibration is limited within a "reasonable" range based on field and modeling experience for a range of similar conditions.

The residual differences between observed field data and the calibrated model represent how well the model fits the field data and provide a form of model test. Calibration residuals are a weak form of model test because the modeler had an opportunity to fit or adjust the parameter values to these data. Thus, when the number of parameters in the model is large or similar to the number of field observations, the utility of calibration residuals for model testing can be small.

Model Testing and Evaluation

Model testing can consist of a wide variety of techniques intended to evaluate and demonstrate the strengths and limitations of a model for particular purposes (Gass 1983, Kleijnen 1995, Beck 2002, Parker et al. 2002). Ideally, model-testing procedures and protocols are established early in the modeling process (Kauffman et al. 2001). Some common forms of model testing include the following items.

Software Tests

Software tests can occur at several levels and by several means (Kauffman et al. 2001). Parts or components of the model can be tested separately, in functional units, and then together as a modeling system. These code tests ideally are done by people other than the authors and can be done by a designated "librarian," a peer-review process, parallel development teams, or a formal individual or group "walk-through" of the code (Ropella et al. 2002, Grimm and Railsback 2005). When programmers understand that others will inspect and test their code, coding tends to be more reliable.

Numerical Tests

Numerical tests are used to ensure that the model's calculations are stable and correct for some well-known cases and solutions. Complex models can be numerically unstable for some cases, and numerical tests can help establish the limits (Sobey 2001). Routine model applications of common software often rely on software and numerical tests done by the model developer and prior applications of the model.

Empirical Tests

Comparisons with field data at the component or system scales are useful tests of a model. Such tests are stronger if they are done with data sets different from those used for model calibration and over a wide range of field conditions (wet and dry years, for example). Unfortunately, field data often are sparse and unavailable for complete empirical testing over a wide range of conditions. Such empirical tests against independent field data often are called "model validation" studies, but the sparseness of field data usually means that such tests do not fully demonstrate the "validity" of the model for all relevant field conditions. Empirical model testing never is directly available for model applications for nonexisting conditions, such as conditions in the future with alternative solutions (Gass 1983). An additional problem is the quality of field data; difficulties and errors in field observations make empirical tests of a model less accurate.

Model Comparison Tests

A large system model often must simplify components or the overall representation of a system relative to detailed models that might exist of the system or system components. Where the detailed model or model components provide greater confidence in the representation (sometimes they do not), then comparison between the complex and simplified models can provide some insights and understanding of the relative limitations of the two models. Model comparisons often can be made over a wide range of virtual field conditions, thus avoiding the limitations and expense of comparisons of model and field results. However, model comparisons are weaker tests than good empirical tests. Model-comparison results also often are used to assess the numerical errors in the model solution method.

Sensitivity Analysis

Sensitivity studies quantify the effects of small changes in model assumptions on model results. Such sensitivity results provide insights into the probable range of error in model results from such causes. Sensitivity results can be useful for interpreting model results and assessing the data quality needed or desirable from field investigations (Rose 1989, Drechsler 1998, Saltelli et al. 2000, Frey and Patil 2002).

Expert Evaluation

Almost all model results are evaluated by experts in the problem being modeled. Such expert evaluation occurs in model development, calibration, and application. Errors are frequent in modeling complex systems, and expert inspections are often the most readily available and capable means to identify potential errors. Expert review commonly is done internally by the modeling team through both informal and structured processes. Additional review by local or external experts on the general type of problem of modeling also can be used.

Overall, as noted by Quade (1980), "a particularly dangerous myth is the belief that a policy model can be fully validated—that is, proved correct. Such models can, at best, be invalidated . . . . Thus the aim of the validation [testing] (or rather invalidation) attempts is to increase the degree of confidence that the events inferred from the model will, in fact, occur under the conditions assumed. When you have tried all the reasonable invalidation procedures you can think of, you will not, of course, have a valid model (and you may not have a model at all). You will, however, have a good understanding of the strengths and weaknesses of the model, and you are able to meet criticisms of omissions by being able to say why something was left out and what difference including it would have made. Knowing the limits of the model's predictive capabilities will enable you to express proper confidence in the results obtained from it."

Every decision maker has a mental model or understanding of the problem (Gass 1983). However, these mental models are tested only indirectly by political election, appointment, or promotion processes that place an individual in a decision-making capacity. It should be possible for quantitative models based on scientific and technical information to demonstrate greater levels of credibility to supplement, aid, or improve on decision makers' mental models and ultimately improve the consideration and selection of decisions.

Interpretation and Communication of Results

Even a perfect model will be useless if its results are not trusted and used for understanding or solving a problem. Model results and their implications must be interpreted and communicated for nonspecialists in the context of the problem. The communication of results must often address two issues: communication and support of insights and results and demonstration of the credibility and limitations of the model and its results. Communication of insights from the results, along with the general degree of confidence in them, often is all that can be provided to busy decision makers. However, the model and its results must also be presented and documented in a form that allows technical workers to understand them more deeply. The formal write-up of the model and its results should aid the clarity and depth of the presentation.

Documentation and External Review

Documentation facilitates training of model users, supports the credibility and transparency of a model (allowing the work to be externally reviewed), and furthers the education of water managers and modelers regarding the problem being modeled. Documentation also has an important internal quality-control function. Documenting a model and the thought that goes into documentation helps to ensure that a model works, so that its limitations are understood and can be communicated, and future improvements can be identified.

Peer or external review can be useful for communicating and establishing model credibility. Such reviews always provide some technical value for an ongoing modeling effort. The mere expectation of external review can lead to improvement in the technical discipline and presentation of modeling. However, a credible model review will almost always find some real or potential flaws, so in an adversarial environment, external reviews can be risky. External review can be conducted in stages throughout the modeling process, at the end of model development, or for specific model applications. External reviews usually are more useful if they are integrated into modeling and application of the model. Although the review process takes time and resources that might have been devoted to additional modeling, at least some level of external review is important for quality and credibility.

Establishing Model Credibility

A primary aspect of model development, testing, and application is establishing the credibility of the modeling effort (Gass 1983). Credibility can be based on

- A model's agreement with specialist or popular notions regarding the system (face validity)
- Credentials of the modeler or modeling organization
- Technical procedures and protocols followed in model development

- Model documentation produced
- Tests conducted on the model and its results
- Qualifications of advisers or reviewers of the effort
- Outcomes of formal external (peer) review
- A (long) period over which the model has been used
- Current model use
- Diversity of situations for which the model has been used
- Authoritative (agency) sponsorship of the model or modeling effort

Some of these factors that bolster the perceived credibility of a model may have little to do with its actual technical reliability, but the wide perception that a model is credible is required for its results to be trusted.

Models developed for applications in an adversarial environment must be pursued with particular care. When a model or its results are expected to enter into legal or political proceedings, an especially systematic, tested, transparent, and articulate modeling effort is required, or an especially astute follow-up and clarification is required after the results are released.

No amount of effort can ensure that a model is perfect. However, following the systematic model-development and application processes described above can greatly increase the likelihood that a model will be useful for understanding or developing solutions for problems.

INSTITUTIONAL MODELS FOR INTEGRATING KNOWLEDGE AND MANAGEMENT

The purpose of applied quantitative modeling for ecosystem management is to provide information and insights to individuals and groups with decision-making and management responsibilities (Geoffrion 1976). These decision makers are in (sometimes competing) institutions that make and support decisions and operations. The purpose of these results and insights is to improve decisions and provide decision makers with greater confidence in the likely effectiveness of their decisions.

Modeling and model results can enter decision making in several ways.

- Directly determine a decision. Direct adoption of solutions suggested by a model is rare. In a few narrow cases, such as selecting the operation of particular hydropower turbines over short periods, model results directly determine decisions. A few water-distribution systems also are operated largely by model results over short periods, mostly as an aid to system operators.
- Provide technical support, along with monitoring data and experi-

ence, for operating decisions. Such support is common for the operation of most large water systems. One or more computer models will be tailored to provide specific information to system operators and managers for hourly, daily, monthly, and longer-term operational decisions. Models provide an ability to estimate field outcomes for locations and times when data are unavailable (such as the future) and provide a timely and less-expensive way to explore operational scenarios under a variety of conditions.

• Provide a major direct part of the negotiating and decision-making environment. Especially for routine technical decision-making, model use is common. Models can be tailored for such situations and provide results, which, although imperfect, provide consistent and insightful results for decision makers experienced with a routine problem. For nonroutine decision making, where conflicts are more common and models are less well-tailored to the problem, models have less of a role. The use of models in negotiations is discussed in more detail below and has sometimes been successful.

• Model results can inform the background for decision making and the decision issue. More commonly, model results provide background information for decision makers, much as any background technical study provides useful information.

• Decision makers, their staffs, and ultimately the public are educated in general through use of models and model results over long periods. For most major water systems, agency staff become educated through development and use of models as well as through direct experience with the system. In the course of such exercises, staff develops an understanding of how the system would perform under a wider variety of circumstances than have been directly experienced. Staff also becomes familiar with the models and their strengths and weaknesses. Modeling staff members often are promoted to middle or senior management, where their reliance on models is less direct but was foundational for their understanding of the system.

The design and execution of modeling efforts should consider the decision-making environment that they are intended to inform. Several kinds of decision-making environments and their implications for modeling are discussed below.

Technical Decision Making

A basic difference exists in the use of models in technical and adversarial discourse and decision making. Technical and scientific decision making ideally examines a wide range of solutions, eliminating those whose performance is unpromising until a final small set of promising alternatives remains, from which one is chosen. There always is uncertainty in all but the most fundamental knowledge (such as conservation of mass, energy, and momentum). As most applied models are based on assumptions beyond fundamental knowledge (empirical knowledge and often professional judgment), almost all models are imperfect and will err in some manner. There is no such thing as a "scientifically valid" model unless it is based on fundamental principles (Konikow and Bredehoeft 1992). Like other scientific hypotheses, a model can only be invalidated. A model can never be completely validated. In an applied context, all model results must be interpreted and judged. Uncertainty always exists (Oreskes 2003).

When there is consensus on objectives, technical and scientific decision making is quite successful. Quantitative models are routinely used and trusted for major water and environmental decisions every day. National Weather Service models of storms and floods have been tremendously effective for reducing loss of lives and property from storms, even though they are imperfect and their results have significant uncertainties. Outside of environmental applications, quantitative models are relied on for increasing the reliability of buildings and bridges, increasing the efficiency of airline schedules, and countless other practical applications. All of these models retain important uncertainties, but they do provide insights and a logical basis for conclusions without which decision making would be more difficult.

Adversarial Decision Making

Where conflict on objectives exists, typical decision-making processes ask more of quantitative models.

Adversarial decision making, which dominates legal and political discourse, is a contest among alternatives or for and against a particular proposed alternative. In such a contest, models and model results supporting an alternative are presented by proponents. Proponents attack or discredit models and model results that do not support the proposition. Adversaries to a particular proposal take the opposite view. In such contests, an uncertain model or imperfect model results are often easily discredited. Adversarial decision making has difficulty in using models and model results without a preponderance of scientific support (Jackson 2006). Communication of model results is especially important in an adversarial process.

In adversarial environments, proponents of the status quo will often call for additional study, detailed modeling, or long periods of data collection, particularly if the models are financed by their opponents. For proponents of the status quo, more studies and modeling are always needed. One of the more productive uses of modeling in adversarial situations is to help reshape understanding of a problem and solutions over a long period. This use poses little urgent threat to the status quo position and allows improved understanding and solutions to be crafted for a future time when the political environment is more fluid, as in the aftermath of a major drought, flood, or lawsuit.

Negotiations

Use of models and model results often is proposed as part of negotiated solutions. The original adaptive management (Holling 1978), "shared vision modeling" (Lund and Palmer 1997, Palmer et al. 1999), and gaming approaches all have in common the use of computer models to represent tradeoffs, certainties, and uncertainties in negotiations among conflicting parties. This decision-making environment lies between pure technical-scientific and adversarial decision making. Where there is broad motivation to come to a consensus agreement and realization that technical support is needed for such an agreement, then models can have a useful, even central, role in negotiated decision making.

Quantitative models can have several roles in policy negotiations:

• *A decision-support system for negotiations.* Here, computer models form the central venue and technical arbiter for negotiations, constituting a substantially agreed-on technical basis for discussions and comparison of performance for proposed or crafted alternatives. Typically a "neutral" technical and scientific party creates the model support for negotiations or a process in which technical representatives from major stakeholders come to an agreement on a model representation of the problem.

• *Model results used directly in negotiations.* Here, model results are used in negotiations, as any technical study or document would be used. This approach does not require as much consensus on the technical merits of the work, and allows the modeling to have a more peripheral role in the negotiation deliberations.

• *Preparation for negotiations.* Models and modeling results often are used in preparation for negotiations. Each party can perform internal modeling studies to investigate options from their perspective and those of other parties to the negotiations. These investigations can help to form the basis of proposals and critiques offered during the negotiation process. Sometimes such internal modeling studies are performed during the course of negotiations.

• *Models used to train technical advisors in negotiations.* Actual negotiations often are on time frames too short for new modeling studies to be done. In such cases, past model studies, often accumulated over decades, provide negotiators or technical advisors to negotiators with considerable knowledge of promising and unpromising alternatives, as well as insights and concerns worthwhile during negotiations.

An adversarial process often follows such a period of negotiation. Even if a negotiation leads to a formal agreement, there are opportunities for further negotiation and adversarial decision making in the implementation of any agreement.

Regulatory Environments

Agencies are tasked with promulgating and enforcing environmental and water regulations, and enforcing laws and property rights. In an ideal world, field-monitoring data would be abundant, precise, and accurate. However, field data are imperfect, typically sparse, and unavailable for hypothetical future conditions. Thus, for routine regulatory proceedings, field data often are unavailable or insufficient alone to make permitting or enforcement assessments. In such cases, quantitative models can have two roles. First, models can interpolate or extrapolate from existing field data (which often are used to calibrate the model and establish boundary conditions) and save the agency and the permittee considerable expense and delay for data collection. Models are used to assess the probable environmental or resource effects and the effectiveness of any proposed mitigation actions. These applications all use model results for a regulatory decision.

Another role of modeling is for more formal accounting of environmental effects. Here, the model is effectively designated as an accounting standard, eliminating human assessment. For water-rights allocations, models—however imperfect they might be—are almost the only practical means to assess water availability in a complex system. The use of quantitative models as a basis for TMDLs and TMDL allocations is a more modest example of the model developing into a standardized understanding of a system. To some degree, the automation of model-based accounting can provide greater transparency and predictability of regulatory decisions, as presumably any party can run the model.

The particular type of resource or environmental regulation also can affect the use of quantitative modeling. Where environmental regulation is based on traditional command and control, including specification of required technology, such as specifying particular wastewater treatment processes or so-called best-management practices, routine model use is less important, although models might be useful for determining which technologies should be required. Where regulations specify only a performance standard, regulated parties can use potentially more economical means to achieve the standard, but monitoring or modeling requirements are increased to make the regulations enforceable. For investment in modeling to be worthwhile compared with investment in monitoring, monitoring must be relatively expensive, and models of performance must be relatively good. For market-based regulations (such as water markets or markets for

TMDL), the use of models as an accounting mechanism becomes attractive because it often is an onerous task to have enough density or accuracy of field monitoring to enforce property rights.

CONCLUSIONS

Despite their scientific imperfectability, models have a variety of uses for ecosystem management, including hydrologic, hydraulic, water-quality, habitat, biological, and management models. The development of these models and suites of models should address many technical concerns, including issues of scale, and should follow a systematic process of development and application, including testing. Model development and application also should be tailored to specific management purposes and decision-making contexts.

Despite their potential—and often-realized—usefulness in decision making, not all models or modeling efforts help to solve the problems they are applied to. The systematic process of development and application referred to above needs to take serious account of the appropriate potential applications, utility, and limitations of the models being considered. As a result, the modeling process itself may or may not help with achievement of stated purposes. This point leads to the committee's discussion in Chapter 6 of the need for integrated management systems and efforts, because even the best models and the best data will not help informed decisions to be made unless the right questions are asked about the performance of the entire system and how the separate components influence that system performance.

4

Natural Flow Study

INTRODUCTION

Chapter 2 described the Klamath River basin as the locale for applying models used by the U.S. Bureau of Reclamation (USBR), and Chapter 3 provided a general introduction to models. This chapter explores a specific model developed by the USBR Natural Flow Study (NFS) (USBR 2005) for estimating flows for the Klamath River at Keno (Figure 4-1).

The following pages introduce the NFS (USBR 2005) by explaining the background and objectives of the model and reviewing its history. The committee assesses the resulting study and its models by addressing the following issues:

- Specific methods used to estimate stream flow, groundwater inflows to Upper Klamath Lake, evapotranspiration, lake levels, and the derivation of Keno gauge discharge from Link River flows using regression analysis.
- Data, including the representativeness of the period of record, quality assurance and quality control (QA/QC), and information gaps.
- Sensitivity, uncertainty, and error propagation.
- Desirable analyses, alternative approaches, and follow-up.
- Scope, context of modeling objectives and strategy, and integration into a larger plan.

Conclusions and recommendations complete the chapter, including a discussion of management implications of the NFS.

FIGURE 4-1 Keno Dam rests atop a bedrock sill across the Klamath River near Keno, Oregon. The fish ladder for the dam is on the left abutment of the dam in the foreground of this view.
SOURCE: Photograph by W.L. Graf, University of South Carolina, July 2006.

Background and Objectives of the Natural Flow Study

The USBR conducted the NFS to "estimate the effects of agricultural development on natural flows in the upper Klamath River basin" using an "estimate of the monthly natural flows in the upper Klamath River at Keno" (USBR 2005, pp. i, ix, 1). Essentially, the USBR study would provide flow estimates that would be observed if *there were no agricultural development such as draining of marshes and diversions of flow* in the upper Klamath basin. The following section reviews the history of the NFS, explains its relationship to Hardy et al. (2006a; also referred to here as the Instream Flow Study [IFS]), and details this committee's interactions with the authors of the study.

History of the Natural Flow Study

J. Hicks (USBR Klamath Project Area Office, personal communication, 2007) provided a useful review of much of the following history of the NFS.

In 1999, Hardy completed a first-phase report containing recommendations for interim instream flows for the main stem of the Klamath River below Iron Gate Dam (Hardy 1999) (Figure 4-2). The purpose of the report was to support the ecological needs of aquatic resources, particularly anadromous fish species. During the early 1990s, water users and managers were entangled in disputes about the appropriation of Klamath River waters. To address this debate, the U.S. Department of Justice contracted with Utah State University (USU) in 1996 to collect information for use in the Klamath Basin Adjudication Alternative Dispute Resolution process. The Bureau of Indian Affairs (BIA) provided the funding. The following paragraphs outline the history of the NFS as determined by the committee's hearings, conversations with participants and observers, input from J. Hicks (USBR Klamath Project Area Office, personal communication, 2007), and communications from T. Hardy, of Utah State University, the principal author of the IFS. Committee members visited the USBR in Denver to collect further information to support this history, which is partly summarized

FIGURE 4-2 The U.S. Geological Survey stream gauge site on the Klamath River immediately below Iron Gate Dam is the site for calculated natural flows and recommended instream flows. The gauge site includes a cable car for sampling and a rectangular housing for data recorder and transmitter.
SOURCE: Photograph by W.L. Graf, University of South Carolina, July 2006.

in Figure 1-4. Those conversations included T. Perry, a hydrologist at the USBR and the principal author of the NFS.

An important aspect of the USU investigation was the estimation of unimpaired Klamath River flows, or those flows that would exist if water storage and diversion for agriculture and national wildlife refuges did not take place. The USU team proposed to estimate unimpaired flows in the Klamath River by adding together irrigation uses, consumptive use in the marshes of the wildlife refuge, and additions to flow below Upper Klamath Lake. The USU acquired consumptive-use estimates for agricultural lands above Upper Klamath Lake from the Oregon Water Resources Department and Upper Klamath Lake net inflow data and flow accretion data below Upper Klamath Lake from THE USBR. The USU team then hired contractor Phillip Williams and Associates (PWA) to use a water-routing model (MIKE 11) to estimate outflows from Upper Klamath Lake without any diversions. PWA was at the same time under contract with the USBR to evaluate the impacts of Upper Klamath Lake level regulation on water quality and endangered suckers.

Also during this time, the USBR consulted with the Fish and Wildlife Service (FWS) and the National Marine Fisheries Service (NMFS) of the National Oceanic and Atmospheric Administration (NOAA) on the Klamath Project's effects on ESA listed species. Since the NMFS had little information on coho in the Klamath River, it relied heavily on the Hardy Phase II Interim Instream Flow Needs report (Hardy and Addley 2001) recommendations. Implementation of the schedules for Upper Klamath Lake surface elevations and Iron Gate Dam release schedules contained in the 2001 biological opinions resulted in the curtailment of water deliveries to the Klamath Project. Project water users suffered significant economic losses and the Klamath Project became the subject of national media.

In 2001, the BIA requested that the USBR fund a revision of the PWA original hydrodynamic modeling for use in the Interim IFS. The USBR contracted with PWA to develop a relatively simple model to estimate historic undepleted flows for the upper Klamath River. The model PWA developed estimated the potential upper Klamath River flow based on the elimination of all agricultural depletions in the upper Klamath River basin. The model did not consider the historical size of Upper Klamath Lake, the effect of its associated marshlands, or the effects of Lower Klamath Lake or the Lost River Slough on historical river flows. The PWA group submitted its report on September 5, 2001, and THE USBR then provided it to USU.

In early 2002, Oregon Water Resources Department commented to USU that the use of their consumptive-use estimates was prohibited, because the data were only to be used for the Alternative Dispute Resolution process. The BIA then requested that the USBR obtain independent consumptive use estimates. The USBR Technical Service Center (TSC) completed this project on October 10, 2002, and provided it to PWA, which finished its modeling on October 21, 2002. The results of PWA's simplified river flow estimates were used in the development of instream flow recommendations for the Interim Instream Flow report.

On November 6, 2002, T. Perry of the USBR TSC sent a Technical Memorandum to the Klamath Basin Area Office manager that identified some of the weaknesses in the simplified PWA modeling exercise (Perry et al. 2002). The memorandum also established a framework for evaluating the necessary components for an actual natural flow study. The TSC recognized that a more complex study of pre-agricultural development would be necessary to estimate more accurately the actual effects of agricultural development on historical upper Klamath River flows.

In late 2002, TSC initiated an intensive study to understand these agricultural effects and completed a draft in December 2004. The report was reviewed by a number of stakeholders, including agency personnel. Comments were provided on the original draft; many were implemented, along with additional model runs, for inclusion in the November 2005 final report (USBR 2005).

The USBR reconsulted with the NMFS and received two new biological opinions in the spring of 2002; the new opinions again contained requirements for stream flows and lake elevations that could result in shortages to the Klamath Project. The NMFS opinions also contained a requirement for acquisition of additional water for release to the river in amounts that increased yearly until they reached 100,000 acre-ft. For comparison, the Klamath Project's average annual consumptive use is approximately 300,000 acre-ft. Annual expenditures to acquire this water have averaged $5.5 million. The large increase in the USBR's funding needs generated numerous questions from government budget officials, members of Congress, and others. One of the most important issues was how the biological opinion requirements compared with pre-project hydrology.

The USBR intended the NFS to be a more detailed study of natural flow of the upper Klamath River than the PWA consultants' study. Neither the NFS nor the Interim IFS was designed to be used in the Endangered Species Act (ESA) consultation; however, the timing of the studies and the consultations may explain the assumption made by stakeholders that the USBR embarked upon the NFS for use in the second consultation with the NOAA.

As federal officials and stakeholders became aware of the NFS, they recognized that it presented a more accurate estimate of pre-agricultural flows in the upper Klamath basin, and that it could be used in the ongoing interim instream flow work. The USBR officials thus began to redirect the focus and intent of the NFS such that it more closely coordinated with the IFS. Hardy et al. (2006a) were especially interested in the quality of the NFS results so that they could determine whether those results were better than those of the PWA study. The instream flow team wanted hourly or at least daily flow estimates, but the USBR informed them that the data required for estimating such flows with a satisfactory degree of accuracy were unavailable and did not provide the data.

The instream flow technical team, which met frequently in Arcata, California, interacted with USBR personnel who conducted working sessions with team members to acquaint them with the methods used in the NFS. Members of the instream flow team suggested to the USBR that the NFS should encompass the entire river, not just the reach above Keno Dam. Team members were dissatisfied when the USBR informed them that this expansion was not possible because of time constraints.

The USBR later agreed to provide natural-flow data as far downstream as Iron Gate Dam in the form of a spreadsheet showing additions to flow between the two dams, even though Iron Gate Dam is below Keno Dam, the cut-off point for the NFS. In addition, the USBR provided the interim flow team with a spreadsheet containing the estimated historical diversions to the Rogue Valley from Jenny Creek (a tributary above Iron Gate Dam).

The combined spreadsheets represented the best available estimates of natural, undepleted additions to flow between Keno and Iron Gate dams. The only depletion from that stretch of the river that was inadvertently omitted was a diversion of 2 cubic feet per second (cfs) to 8 cfs to the City of Yreka. The instream flow team elected to use a simplified estimate of accretion rather than the data provided by U.S. Geological Survey (USGS) data and estimated diversions. The USBR noted that the flows the team chose to use were considerably higher than those provided by the USBR.

When USBR staff asked the instream flow team which data they were going to use in their model for natural flow estimates for the rest of the Klamath River watershed, the team replied that it had only the USGS gauge data for impaired flows and intended to use the impaired flow data. USBR staff believed it would be inappropriate to use natural flows in the model for only one section of the river and impaired flows for the rest of the watershed. Recognizing that the instream flow team was trying to meet a deadline, USBR staff quickly estimated unimpaired flows for the Shasta, Scott, and Trinity rivers as well as some of the major creeks, and provided those data.

The foregoing brief review of the origins of the NFS demonstrates that a variety of administrative forces were at work in the creation of the study. Biological opinions and the need for improved understanding of the ecological characteristics of the hydrologic system became imperative, but only after the initiation of the study. The study came to be an outgrowth of more than one objective and method, increasing the difficulty in creating a useful product. Using the output of the NFS as input to the IFS seems now, after the fact, to be a logical thing to do, but because this intended use was not apparent at the beginning of the entire process, shortcomings in such a connection were likely. As outlined later in this chapter, the final product of the NFS did make some contributions to scientific understanding of the hydrology of the Klamath River. There were also some shortcomings that might have been avoided if there had been greater coordination among hydrologic, ecological, and operations researchers at the beginning of the NFS.

Meeting with the National Research Council Committee

A group of members of the NRC committee and NRC staff met with USBR staff in Denver on November 20, 2006. Thomas Perry, USBR Technical Lead, Jon Hicks, of the USBR Klamath Basin Area Office (KBAO), and other USBR personnel were also present. Perry's PowerPoint presentation and comments provided much of the information to the group.

During Perry's PowerPoint presentation, he emphasized that the prime directive to the NFS team was "do not underestimate natural flows." During his discussion of the sensitivity analysis, he showed that the model produces an increase in flow at the Keno gauge when the consumptive use in the upper Klamath basin is increased. In other words, more water used by agriculture means more water going into the Klamath River. Perry posited that this phenomenon had something to do with "water limiting" conditions for evapotranspiration in Upper Klamath Lake. This proposition is discussed in a subsequent section of this chapter.

Natural Flow Study General Approach

The USBR used a water-budget approach to estimate natural flows at the Keno gauge. As stated by USBR (2005, p. ix):

> The approach was to evaluate the changes of agriculture from predevelopment depletions, estimate the effects of these changes, and restore the water budget to natural conditions by reversing the effects of agricultural development. Records used in this empirical assessment were derived from both stream gaging flow histories and from climatological records for stations within and adjacent to the study area.

The emphasis was on the effects of agricultural development; other changes, such as changes in forest cover, were not assessed (USBR 2005, p. xiii). The USBR first developed a reasonably accurate conceptual model (Figure 4-3) to identify all the significant components of the basin water budget. The model was useful for identification and the subsequent quan-

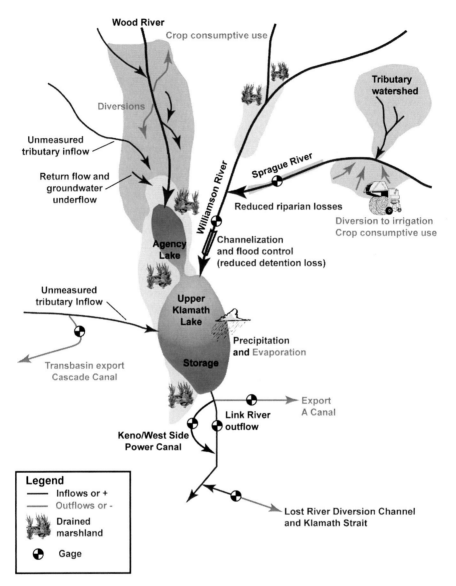

FIGURE 4-3 Conceptual model of the NFS.
SOURCE: USBR 2005.

tification of gains and losses associated with the process of "naturalizing" the current conditions.

The NFS produced results of the water-budget approach given as monthly flows at the Link River gauge and the Keno gauge on the Klamath River. The Keno gauge flows were estimated from a regression equation between the Link River gauge and the Keno gauge. The period of record was 52 years, from 1949 through 2000. A longer period was not included because of the generally unreliable data prior to 1949 and the lack of any data at the start of the study that was newer than 2000 (T. Perry, USBR, personal communication, 2006).

The NFS used the following generalized steps (USBR 2005):

- Compute naturalized flows from the major tributaries—the Sprague, Williamson, and Wood rivers—to Upper Klamath Lake.
- Compute groundwater inflows into Upper Klamath Lake.
- Develop natural stage-storage and stage-discharge relationships for Upper Klamath Lake.
- Perform a detailed water-budget analysis of Upper Klamath Lake using naturalized inflows, natural (predevelopment) evapotranspiration from marshes surrounding Upper Klamath Lake and the open water surface within Upper Klamath Lake and the stage-storage, stage-discharge relationships to compute naturalized outflows from Upper Klamath Lake to yield naturalized flows at the Link River gauge.
- Develop and use a regression equation to compute naturalized flows at the Keno gauge using naturalized flows at the Link River gauge.

The study implemented the water-budget approach using an Excel spreadsheet and customized computational modules developed specifically for the spreadsheet. Other models were not used because "this study is unique" (USBR 2005, p. xii).

METHODS FOR THE NATURAL FLOW STUDY

Stream Flow

The NFS used a water-budget approach to compute predevelopment flows in the upper Klamath basin. The NFS determined natural flows at the location corresponding to the current Keno gauge by developing a regression equation between Link River and Keno gauges. The basic water-budget approach attempts to "naturalize" flows by adjusting the gauged stream flows to account for losses and gains due to such changes in the basin as agricultural practices and loss of natural marshes along the streams due to watershed development.

Computation of the naturalized flows from the tributary watersheds of Upper Klamath Lake required the adjustment of gauged flows using the following equation:

Natural flow = (gauged flow) + (crop net consumptive use)
  – (reclaimed natural marshland net evapotranspiration).   (4-1)

Although major tributaries have historical records, their records often have missing values. The NFS used a correlation analysis to restore such missing values and create complete gauge records. At other locations where gauged flows are unavailable, techniques of stream flow estimation for ungauged basins have been used. Figure 4-4 is a flowchart of the approach used for naturalizing stream flows.

Groundwater

Subsurface inflow to Upper Klamath Lake is a significant unmeasured component of the lake's water budget (USBR 2005, Attachment E). Assessment of this water-budget component is difficult. The ongoing USGS-Oregon Department of Water Resources study, due to be completed in 2007 or 2008, should help to quantify the uncertainty, which in turn will allow better understanding of the significance of subsurface inflow and its impact on model outcomes. One of the key elements of this study is a groundwater flow model, which can be used to estimate groundwater-lake interactions.

The NFS used estimates developed by Hubbard (1970), who in turn based his estimates on a water budget for Upper Klamath Lake over a 2-year period, 1965-1967. The amount of water required to balance the lake's water budget was an "input" term, which Hubbard assumed to be due to groundwater inflow. This is the "derived groundwater inflow" used by the NFS (USBR 2005, Attachment E). In this approach, one is simply solving for the unknown in the water budget. This approach has the disadvantage that the unknown term, a residual, contains all the cumulative errors. The terms Hubbard assumed to be "known" (calculable) include surface-water inflow, precipitation, evapotranspiration (calculated using the the Blaney-Criddle method); open-water evaporation, and storage. Hubbard used an incorrect relationship for the area capacity of Upper Klamath Lake and for inundated marsh areas, because he incorrectly referenced USGS elevations to the USBR datum. The NFS corrected those two shortcomings.

The corrected USBR estimates of groundwater accrual to Upper Klamath Lake averaged about 19,500 acre-ft per month, assumed to emanate from the regional aquifer. The NFS adjusted the groundwater inflows for a climate signal, based on the inferred climate-influenced discharge from the deep regional aquifer in the region of the lake (Gannett et al. 2003).

In several of the upper Klamath basin watersheds, groundwater pumping is significant, from either the regional confined aquifer (Sprague and Williamson valleys) or the valley-fill alluvial aquifer (Wood River valley). M. Gannett (USGS, personal communication, 2007) reported that in 2000, total pumpage in the upper Klamath basin was about 150,000 acre-ft, or about 10% of the average annual discharge at Iron Gate Dam.

In the case of the regional aquifer, the NFS assumed that the deep

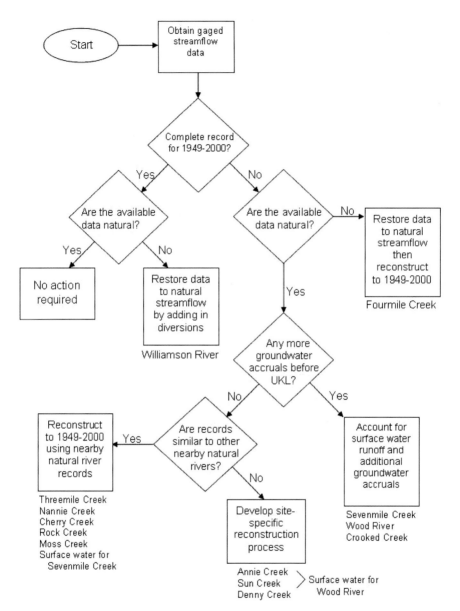

FIGURE 4-4 Flowchart of natural stream flow methods used for computing natural inflows to Upper Klamath Lake.
SOURCE: USBR 2005, Attachment B.

regional aquifer was essentially "disconnected" from the shallow aquifer and so the pumping effect on stream flow was not assessed. Since the Wood River valley wells pump from the shallow valley-fill aquifer, the NFS estimated that about 6,100 acre-ft per year occurred as unmeasured inflow into Upper Klamath Lake. Presumably this figure is included in the figure cited in the previous section. In any case, the number is insignificant given the total water budget of Upper Klamath Lake.

The NFS stated that pumping is significant in several of the upper Klamath basin watersheds, but no indication is given as to how much "significant" is. It therefore is difficult to assess whether the study approximated groundwater effects with any certainty (except for the Wood River valley).

The groundwater inflow to Upper Klamath Lake is estimated as a residual. As the report itself states, this approach leaves much to be desired. The USGS groundwater-lake model was not completed in time for the NFS. In his presentation to the committee, Gannett noted that average groundwater inflow to Upper Klamath Lake is on the order of 320-350 cfs (232,000-253,000 acre-ft/year); these numbers are in general agreement with those in the NFS (USBR 2005, Table E-1). Gannett also provided estimated groundwater discharge to subareas in the upper Klamath basin.

Nevertheless, the NFS made no effort to test its estimate of groundwater inflow. A crude check could have been performed by constructing a contour map of the aquifer's (or aquifers') potentiometric surface(s) in the vicinity of the lake and then doing a calculation with Darcy's Law (Fetter 2001, pp. 122-125) to estimate influx—essentially constructing a flow net (Fetter 2001, Section 4.11). In his Denver presentation, Perry showed slides of the water table elevations surrounding the lake (as did Gannett), but nowhere did the USBR quantitatively use these data. But the problem of identifying which aquifer(s) discharged to the lake and the correct potentiometric surface(s) still would have been present. The method used in the NFS estimate of groundwater inflow appears to be reasonable, especially given the paucity of data.

Evapotranspiration

The NFS used Equation 4-1, the water-budget equation, to estimate monthly naturalized flows for various sub-basins above Upper Klamath Lake. The NFS computed irrigated crop net evapotranspiration values as the net difference between the consumptive use (synonymous with crop evapotranspiration) and effective precipitation, both estimated using the modified Blaney-Criddle method (SCS 1970). The same method also estimated marsh evapotranspiration values, although the USBR adjusted both crop net evapotranspiration values and the marsh evapotranspiration values to mimic the average crop values reported by Cuenca et al. (1992) and some field data for marsh evapotranspiration reported by Bidlake and Payne (USBR 2005, Attachment A, p. 10).

The USBR multiplied estimated net evapotranspiration values by estimates of acreages (crop and marsh acreages as they changed through time) before they were added to gauged stream flow to compute the naturalized flow. For crop net evapotranspiration calculations, the NFS assumed alfalfa, pasture, and spring grain, whereas for marsh evapotranspiration it assumed that tules, cattails, salt grass, and willow could approximate the predevelopment marsh land cover (Figure 4-5). The selection of particular marsh types was constrained by the particular software used for evapotranspiration calculation (XCONSVB) developed by the USBR. The crop coefficients for each crop/land cover type have been derived from Technical Release 21 of SCS (1970). The NFS did not account for diversion losses and return flows, presumably based on a major assumption that those quantities of water would have been included in the gauged stream flow. Furthermore,

FIGURE 4-5 Agriculture is an important economic activity in the upper Klamath basin. This farm, north of Klamath Falls, occupies an area that once was marsh near Upper Klamath Lake.
SOURCE: Photograph by W.L. Graf, University of South Carolina, July 2006.

the NFS considered no water shortages, and consequently, the estimated evapotranspiration values constitute maximum rates under ideal growing conditions (USBR 2005, Attachment A, p. 2).

The approach used in the NFS also required the estimation of evapotranspiration from the open-water surface of Upper Klamath Lake. Considering the unavailability of extensive data required for more accurate evapotranspiration estimation methods, the NFS used the Hargreaves equation (Jensen et al. 1990, Stockholm University 2003), primarily a temperature-based method (Jensen et al. 1990), calibrated to evapotranspiration estimates computed by the Kimberly-Penman method using measured weather parameters at a nearby AgriMet station (USBR 2007a). Lake evapotranspiration estimation also incorporated marsh water limiting functions which assumed a reduction in evapotranspiration as the depth to groundwater increases in marsh areas surrounding the lake (Figure 4-6). It is not clear if these limiting functions have been calibrated in any manner.

FIGURE 4-6 The NFS adjusted flows from the upper Klamath River basin to account for the conversion of marsh areas to agriculture, such as shown here on the southeast side of Upper Klamath Lake. Levees separated the area now under cultivation (on the left) from the lake and its marsh area (on the right).
SOURCE: Photograph by W.L. Graf, University of South Carolina, July 2006.

Net evapotranspiration estimates from crop and marshlands used for naturalization of gauged flows play a major role in the seasonal and inter-annual patterns of the resulting natural flows. Consequently, such flows may have significant uncertainties depending upon the accuracy of various assumptions employed in the naturalization approach. Potential causes of such uncertainties could be due to one or more of the following four aspects of the methods used in the NFS.

The use of the temperature-based Blaney-Criddle method for comput-ing consumptive use raises questions about the validity of the results of the evapotranspiration calculations. Jensen et al. (1990) and Cuenca (1989) have noted the poor performance of the SCS modified Blaney-Criddle method. A major problem of temperature-based evapotranspiration estima-tion methods is the "lag" of peak evapotranspiration estimates compared with the temperature cycle. Cuenca (1989, p. 122) also indicated that the SCS modified Blaney-Criddle method underestimates evapotranspiration, probably due to the averaging of daytime and nighttime temperatures. The exact magnitude of this underestimation for the Klamath region and its implication for the water-budget estimates of the NFS, which depends on the acreages of land cover used in the model, are unknown and likely sig-nificant. Although the SCS modified Blaney-Criddle method is widely used for designing irrigation systems, it is not clear that it can provide reliable estimates of evapotranspiration for computing actual evapotranspiration estimates in watershed modeling.

Cuenca et al. (1992) and field estimates of marsh evapotranspiration by Bidlake and Payne (1998) provided adjustments for the Blaney-Criddle evapotranspiration estimates. The NFS compared the mean net evapotrans-piration estimates computed by the SCS modified Blaney-Criddle method with the median net irrigation values for alfalfa and pasture as reported in Cuenca et al. (1992). Based on this comparison, alfalfa values were not adjusted since they compared reasonably well. However, SCS modified Blaney-Criddle estimates for pasture were lower than those reported by Cuenca et al. (1992), and adjustment factors were computed for the months of April through October. There was no verification that the use of the fixed adjustment factors based on the comparison of average (or median in case of Cuenca et al. 1992) for the entire time period of evapotranspiration estimates will correct any underestimation by the SCS method under all climatic conditions. The same fact is true for adjustment of marsh evapo-transpiration estimates using limited data for wetlands reported by Bidlake and Payne (1998).

It is not clear why the IFS team compared and adjusted net evapotrans-piration estimates (evapotranspiration minus effective precipitation), rather than using the raw evapotranspiration values. Clearly, the adjustment of raw evapotranspiration values before effective precipitation is subtracted may result in a totally different set of net evapotranspiration values. The implications of this choice for the naturalized flows are not clear.

One data set that is available is the evapotranspiration record computed using real weather data measured at the AgriMet station KFLO (Klamath Falls). AgriMet (USBR 2007a) evapotranspiration values are computed using the Kimberly-Penman equation and are available for alfalfa as the reference crop. The comparison of evapotranspiration values reported in Cuenca et al. (1992) with AgriMet data shown in Figure 4-7 indicates the actual evapotranspiration may have been underestimated particularly during peak growth stages when this comparison is likely to be more appropriate. A similar comparison can probably be made for pasture.

Calendar Month

An additional issue in assessing the evapotranspiration calculations concerns irrigation efficiency. In a typical situation, inefficiencies in the irrigation systems result in additional water to be delivered at the diversion point of the stream to account for net losses in delivering water from the streams to crops. Jensen et al. (1990, p. 73) provided the definition of

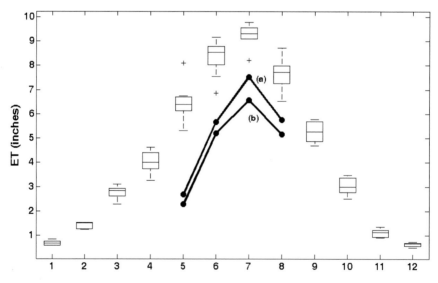

FIGURE 4-7 Comparison of Kimberly-Penman evapotranspiration estimates for alfalfa reported for KFLO AgriMet station (box-whisker plots) with the evapotranspiration estimates of Cuenca et al. (1992, p.128): (a) evapotranspiration values corresponding to 19 out of 20 years; and (b) evapotranspiration values corresponding to 5 out of 10 years.

several efficiency factors that needed to be considered: reservoir storage efficiency (if applicable), conveyance efficiency, and unit-irrigation efficiency. Cuenca (1989) provided other definitions. The NFS likely discounted the use of the irrigation efficiency under the assumption that such losses due to inefficiency will appear at the gauged stream location, and therefore, adding more quantities to consumptive use is not necessary. The assumption that conveyance and storage losses in totality will appear at the gauge location is not easily verified and probably invalid. Any underestimation of the actual net irrigation will underestimate the natural flow although the exact impact of this assumption on exact magnitude of the resulting natural flow is not clear.

In the approach used for naturalizing gauged flows, the NFS multiplied the computed monthly net evapotranspiration values by temporally varying acreages to estimate the total consumptive use in acre-feet (USBR 2005, Attachment A, p. 5). Because proper records of agricultural use and land-cover variations over the 40-year (1961-2000) period of simulation may not be available, the assumptions used to estimate the temporal function of acreages may have significant uncertainties. Actual cropping pattern is influenced by many factors including climate, local economy, and ownership. Clearly, errors in the temporal variation of acreages will influence the interannual pattern of naturalized flows. The uncertainty in naturalized flows due to errors in acreages was not discussed in the report.

Upper Klamath Lake Levels

Routing of naturalized inflows through Upper Klamath Lake required the development of a stage-storage relationship for Upper Klamath Lake under natural conditions prior to the construction of the Link River Dam and other changes. This was accomplished by using available historical data (USBR 2005, Attachment F, pp. 1-4). The stage-discharge relationship under uncontrolled outflow conditions from Upper Klamath Lake was derived primarily from the historical outflow data for the period 1904 to 1918. Actual routing of naturalized flows through Upper Klamath Lake was accomplished through the use of the following water-budget equation:

$$\begin{aligned}
S_{t+1} = S_t &+ \text{naturalized inflow from tributaries (t)} \\
&+ \text{groundwater inflow (t)} - \text{marsh ET (t)} \\
&- \text{open water surface ET (t)} \\
&- \text{natural outflow to Link River (t)},
\end{aligned} \qquad (4\text{-}2)$$

where S_t represents the Upper Klamath Lake storage at the beginning of month t, which is a function of lake stage. Natural outflow under uncontrolled conditions is also a function of lake stage. Marsh ET (evapotranspiration) is a function of temporally varying marsh area covered by the lake's water surface and includes a water-limiting condition to decrease evapo-transpiration as the water table depth increases. The NFS used a third-order Runge-Kutta solution technique (Chow et al. 1988) to solve the above equation for Upper Klamath Lake stage under natural conditions. The NFS "calibrated" the actual implementation of this water-budget model for Up-per Klamath Lake by comparing simulated water levels and outflows with the historical values measured during 1906 to 1909.

The NFS claims that that Runge-Kutta method yielded excellent results (USBR 2005, Attachment F, p. 11). The comparison of the computed outfall with the gauged outfall from Upper Klamath Lake as well as the simulated lake elevations with gauged elevations were provided as evidence for the accuracy of the NFS. However, because the measured flows during the early 1900s were used for developing the rating curve under uncontrolled conditions, the comparison of the model results for the same period may not constitute a true test of the routing under natural conditions.

Link-Keno Gauges Regression Analysis

From the end of the Link River reach, the Klamath River flows through Lake Ewauna and then through a low-gradient reach to Keno, where a ba-salt reef (sill) at elevation 4083.1 ft above mean sea level (msl) historically created a backwater upstream. Keno Dam, operated by PacifiCorp, cur-rently serves a similar function in roughly the same location as the historical reef. During high pre-dam flows from Upper Klamath Lake, water levels upstream of Keno would rise and essentially the entire reach from Keno upstream to the base of Link River would be a long lake.

This lake-like reach of the Klamath River connected with Lower Klam-ath Lake through extensive marshes. In effect, during high runoff, the Klamath River backed up above Keno and spread out into Lower Klamath Lake, as described by Weddell (2000) and NRC (2004a). Storage of flood-waters in Lower Klamath Lake had important effects on the hydrograph of the Klamath River. Lower Klamath Lake functioned as a flood plain, absorbing high flows and releasing them slowly back to the stream channel as stage in the river fell and the hydraulic gradient reversed. Flood peaks were attenuated by this flood storage, and base flows were augmented. In addition, at sufficiently high stage, some water, as much as 12,000 cfs, flowed from Lake Ewauna via Lost River Slough into the Lost River and was lost from the Klamath River system. Under current management, some floodwaters can be diverted into the Lost River system.

In 1907, the California Northeastern Railway Company constructed a railway on an earthen embankment that served to prevent overflow from the Klamath River into the Lower Klamath Lake basin, except under a bridge over the Klamath Straits (USBR 2000). The railroad entered into an agreement with the U.S. government to build and maintain the railroad to serve as a levee, which can be considered a part of the effects of the Klamath Project. In 1917, Klamath Strait was fitted with a gate so it could be closed off, eliminating the hydrologic connection between the river and Lower Klamath Lake. Thus, as of 1907, the hydrologic connection between river and Lower Klamath Lake was reduced, and by 1917, it was essentially eliminated.

One can infer that severing the hydrologic connection from the river to Lower Klamath Lake had significant effects on the flow regime from Keno downstream. By eliminating the flood storage formerly provided by the lake, one would expect flood peaks delivered to the channel downstream of Keno to have increased and the recession limbs to have become less gradual in the absence of "return flow" from water stored in Lower Klamath Lake. From a water-quality perspective, one can infer some changes because this return flow was likely of poor quality, having inundated organic-rich marshland exposed to solar heating. In any event, this historical overflow into Lower Klamath Lake was an important feature of the Klamath River, and its elimination was an important part of the changes effected by the USBR project. Thus, in modeling natural flows at Keno, it would be important to fully investigate this historical hydrologic interaction. This is especially the case if the output from the hydrologic model is to be used as input to a model of physical habitat, as channel form and habitat conditions could respond strongly to changes in flood magnitudes and characteristics of the seasonal recession limb.

In the NFS, the USBR (2005, Attachment F, pp. 15-16) stated,

> Several attempts were made to model this complex system using a digital representation in Excel of the lake/river physical interaction. Because of the complexity of the hydraulics, and the need for detailed data, this modeling is not possible at this time. A correlation approach was used to estimate Keno flows at the outfall of the LKL [Lower Klamath Lake] system. Measured data for the period of October 1904 through September 1918 for the Link River and Keno gauges were used.

Given the likely importance of the historical interaction between the Klamath River channel and Lower Klamath Lake, it is surprising that more effort was not invested in developing a model of the physical interactions. Certainly, detailed topography could be obtained. The USBR (2005) did not explain what "detailed data" were unavailable and why the hydraulics would be prohibitively complex. It may be that to model channel-Lower Klamath Lake interactions would require using a weekly or daily time step, which would be more relevant for the ecological effects ultimately to be modeled, but if the decision to model at a monthly time step only had already been taken, this might have made a more sophisticated model unworkable.

The correlation was evidently done with mean monthly flows only.

Given that flows into and out of Lower Klamath Lake would have responded to differences in relative stage on a much shorter time step, the choice of monthly time step is questionable for the model purpose.

The USBR (2005) ran correlations separately for the high-flow period November through June and another for the low-flow period July through October. In undertaking a correlation approach to analyze this phenomenon, presumably one would want to analyze the data for periods when the Klamath River was overflowing into Lower Klamath Lake separately from periods when water stored in Lower Klamath Lake was draining back into the river channel. This could best be done using the water-surface stage, that is, an if-then statement could be written to sort the data for times when stage at Ewauna or Keno exceeded 4,085 ft msl (or whatever elevation is most appropriate) when the river was flowing into Lower Klamath Lake. The time step is critical here, as the period of overflow into Lower Klamath Lake in many years may have been measured in weeks not months, so that by using a monthly rather than weekly or daily time step, important information could be lost. The USBR (2005) did not present an analysis demonstrating that the monthly time step was appropriate, nor that June 30 divided high from low-flow periods in most years, nor that the seasonal patterns were so consistent that a rigid, fixed date was appropriate instead of using the actual stage data, which were readily available and would appear to be most suited to this modeling.

The NFS ran a correlation for years 1904-1918 except for October 1909 through December 1911 when the shape of the Link River hydrograph was "of special interest" (USBR 2005, Attachment F, p. 17). This is a period of high flows recorded at the Link River gauge with much lower flows at Keno, but during a period when the gauge was subject to backwater effects from logjams, so the USBR (2005) attributes the high flows recorded at Link River to gauging errors. It always is difficult to deal with potentially inaccurate data, so this decision is a reasonable one.

However, the correlation analysis otherwise treated the entire period 1904-1918 as being the same, despite the fact that the hydrologic connectivity between channel and Lower Klamath Lake was reduced in 1907 and essentially eliminated in 1917. Thus, one could argue that the data should be treated differently pre-levee (1907) and post-closure of Klamath Straits (1917), not lumped together. Ignoring these significant historical changes is not justified in the NFS.

The hydrologic interactions between the river channel and Lower Klamath Lake were probably of enormous importance to peak flows and the seasonal recession limb in the Klamath River from Keno downstream. For such an important interaction, the use of a crude regression model with a monthly time step, using a rigid (and not explicitly justified) division between pre- and post-June 30 (instead of using readily available stage data), and apparently ignoring construction of a berm and gate that would prevent overflow to Lower Klamath Lake, seems an inadequate treatment of an important interaction and its historical modification.

DATA FOR THE NATURAL FLOW STUDY

Representativeness of Climatic Data

The period of record used for the development of the naturalized flows in the upper Klamath basin consists of the 52 years from 1949 to 2000. The basis for the selection of 1949 as the beginning year of the period of analysis was the lack of adequate hydrologic and meteorological data prior to this year. For almost all analyses, the convention of water year (from October 1 through September 30) was used.

Because naturalized flows at Keno have been used in the IFS for the development of instream flows below Iron Gate Dam, the representativeness of the selected period of record with respect to its climatic regime is extremely important. For example, if the selected period is drier than the average, then the resulting instream flows may not be appropriate if the next several decades happen to be wetter than normal. Often the question of representativeness of the period of analysis is complicated by the presence of long-term climatic variations, often decades long. For example, the warm and cold periods of the Atlantic Multidecadal Oscillation (AMO) have been shown to have good correlation with rainfall and river flows in the continental United States (Enfield et al. 2001). The El Niño-Southern Oscillation (ENSO) phenomenon has teleconnections to many parts of the United States and in particular, the Trans-Niño index related to equatorial sea surface temperature has the potential for improving seasonal forecasts for the Klamath basin (Kennedy et al. 2006).

On behalf of the Yurok Tribe, Hecht and Kamman (1996) investigated the long-term climatic regime in the Klamath basin. This investigation used the plots of cumulative deviations of the long-term means of precipitation at Yreka and Klamath Falls to identify periods of drier and wetter periods in the long-term records. In addition, the study analyzed the net inflow to Upper Klamath Lake using the same approach. However, there was no consistency in the long-term spells identified from the precipitation records and the net inflow record (Figure 4-4). Hecht and Kamman (1996) attributed the smoothness and the lesser degree of fluctuation observed in the case of the net inflow record to the damping provided by the groundwater storage capacities, resulting in greater persistence of inter-annual flows in the upper Klamath basin.

Risley et al. (2005) used the Pacific Decadal Oscillation (PDO) as an indicator of long-term climatic trends for seasonal flow forecasting in the

Klamath River basin. Based on the period of record of the PDO index, both wet ("cool") and dry ("warm") periods were identified, and they are shown in Figure 4-8 for comparison with the climatic spells identified by Hecht and Kamman (1996). As seen from Figure 4-8, the previous investigations do not show sufficient information to characterize the period of records used for the NFS as being drier or wetter. However, subperiods of drier and wetter regimes can be identified within the 1949-2000 period.

Individual station records often do not reflect regional climatic patterns. To investigate the long-term regional trends in climatic regime, the committee analyzed the Climate Division precipitation data for the period 1895 to 2005 for both Oregon and California. The upper Klamath basin, including the Williamson and Sprague River basins, falls within the boundary of Climate Division 5 of Oregon, whereas most of the lower Klamath basin is located within Climate Division 1 of California. Figure 4-9 shows the

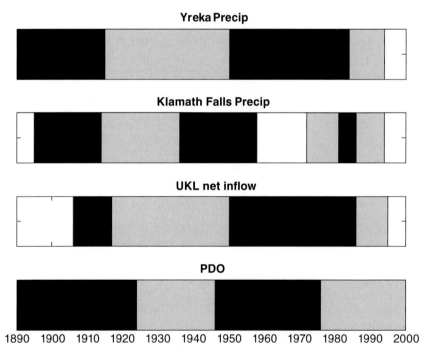

FIGURE 4-8 Climatic spells identified from precipitation records at Yreka and Klamath Falls and from Upper Klamath Lake net inflow time series (Hecht and Kamman 1996). Also shown are the climatic regimes identified from the Pacific Decadal Index (Risley et al. 2005). Dark and light boxes represent wetter and drier regimes, respectively.

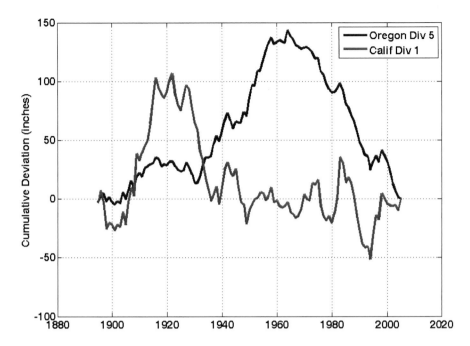

FIGURE 4-9 Cumulative departure from average precipitation for Climate Division 5 of Oregon and Climate Division 1 of California.

cumulative departure of average annual precipitation for these two climate divisions. In this plot, periods of wetter climate are indicated by increasing trends and periods of drier climate are indicated by decreasing trends. The records for these two climate divisions show that there have been distinct shifts in the climate of the upper and lower parts of the Klamath River basin, and these shifts are somewhat out of phase. Two major trends are evident in this figure. First, it appears that starting in the late 1920s, the climate in California Division 1 shifted from a wetter to a drier regime, whereas the climate in Oregon Division 5 shifted from an average to a wetter-than-average regime. This wet climatic episode in the early twentieth century strongly implies that the first two decades of the century are flawed for estimating natural flows, regardless of other changes to hydrology or land use. Then, in the mid-1960s the climate in Oregon Division 5 shifted to a drier regime, and this appears to have persisted to the present. The period of record chosen for the NFS thus includes 16 years of the wetter regime (1949-1964) and 36 years of the drier regime (1965-2000). Consequently, the period of record used in developing estimates of natural flows includes many more dry years than wet years.

Figure 4-10 compares exceedance probability curves of annual precipitation in Oregon Climate Division 5 (upper panel) and California Climate Division 1 (lower panel) for two separate periods, 1895-1948 and 1949-2000. These curves indicate the probability that the average annual precipitation within a climate division will fall below a given value in any year. The upper panel in Figure 4-10 shows that the curve for the period 1949-2000 lies below the curve for the period 1895-1948, indicating that annual precipitation in Oregon Climate Division 5 was generally lower in the latter half of the twentieth century than it was in the first half. However, it also appears that the variability in precipitation in the period was larger than that of the earlier period. The exceedance probability curves for California Climate Division 1 (Figure 4-10b) show that there is little difference in the distributions of annual precipitation for the two periods, suggesting that the 1949-2000 period is representative of the prior climate regime of the lower Klamath basin.

Representativeness of Stream Flow Data

Gauged stream flows are an important component of the water-budget analysis. One key gauge—Williamson River below Sprague River near Chiloquin—has been in operation since 1918. The flow measurements from this gauge are particularly useful because the earliest part of the record spans a period (pre-1950) when development within the upper basin was relatively small (Risley and Laenen, 1999). In addition, the Williamson River contributes roughly 60% of the total inflow to Upper Klamath Lake (percentage estimated on the basis of values given in Figure 4, USBR 2005), thus the measurements at this gauge provide a good indication of long-term changes in the volume of water entering the lake.

Figure 4-11 shows variations in the annual discharge of the Williamson River for the period 1918-2005. This plot indicates that, in the period 1918-1948, the average annual discharge of the Williamson River was generally lower than it was in the period 1949-2000, the base period used in the NFS. The stream flow records for the Williamson River indicate that there was a significant shift in the long-term pattern of annual runoff in the upper Klamath River basin from drier to wetter conditions starting about 1950. This observation is inconsistent with the climate data presented earlier showing that annual precipitation in Oregon Division 5 was generally below average in the period 1949-2000. These inconsistencies most likely reflect limitations in the databases (sparse coverage), not true differences in hydrology. In a previous USGS study, Risley and Laenen (1999) compared precipitation and runoff records from different stations in the region, and came to similar conclusions. Their analysis of flow records showed that most rivers in the regions experienced a clear shift in runoff starting around

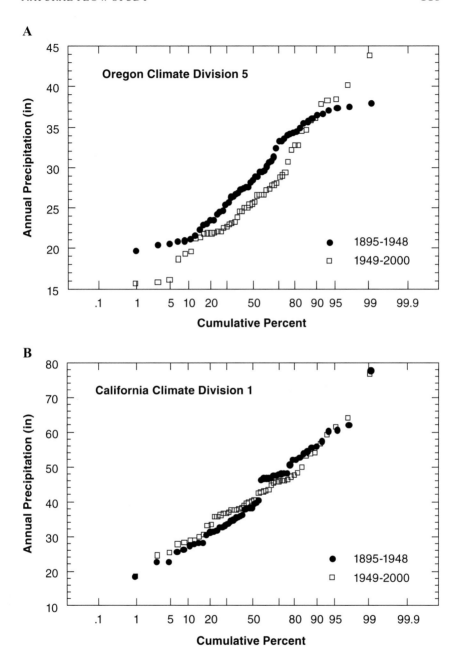

FIGURE 4-10 Annual precipitation data for (A) Climate Division 5 of Oregon and (B) Climate Division 1 of California for the periods 1895-1948 (circles) and 1949-2000 (squares).

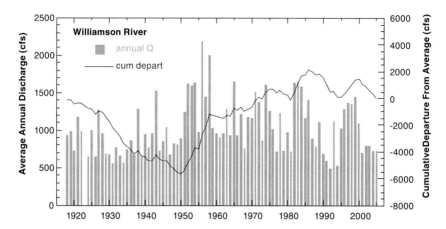

FIGURE 4-11 Average annual stream flow of the Williamson River below the Sprague River near Chiloquin, Oregon, 1918-2005, USGS gauge. Bars indicate the average annual discharge and the line represents the cumulative departure from the average.

1950. However, in similar comparisons of precipitation data, they did not detect significant differences in mean annual precipitation at any of the nearby weather stations (Klamath Falls, Crater Lake, and Medford). Risley and Laenen (1999) provided additional data showing historical trends in timber harvesting and irrigated acreage within the Williamson River basin, but they stopped short of suggesting that land-use changes were responsible for the apparent differences between precipitation and runoff.

The discrepancies between precipitation and stream flow are summarized in Figure 4-12, which compares cumulative departures in annual precipitation for Oregon Climate Division 5 and Crater Lake with cumulative departures in annual discharge for the Williamson River below the Sprague River. This plot reinforces the points made above that precipitation and stream flow appear to be out of phase over much of the period of record. One interesting pattern that emerges from this comparison is that the minimum departure in stream flow (1950) lags the minimum departure in precipitation by about 20 years. Likewise, the maximum departure in stream flow (1985) lags the maximum departure in precipitation by the same amount (about 20 years). The lags between precipitation and runoff could be pure coincidence, and are not likely to be resolved with further analyses of precipitation and stream flow data. However, it seems reasonable to hypothesize that with a clearer understanding of the groundwater system, and the potential role of land-cover changes, it may be possible to

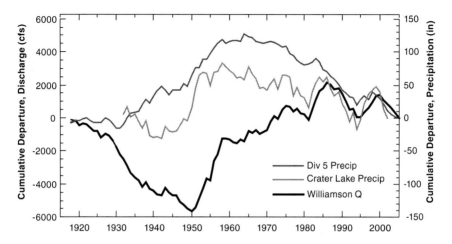

FIGURE 4-12 Cumulative departures in annual precipitation in Oregon Climate Division 5 and at Crater Lake, Oregon, and annual discharge for the Williamson River below the Sprague River.

draw a connection between the differences in precipitation and surface-water hydrology in the upper Klamath River basin.

As noted earlier in this chapter, the naturalized flows from tributaries to Upper Klamath Lake are computed from the sum of three terms: gauged flows, losses due to consumptive use, and gains due to changes in evapotranspiration associated with marshland reclamation. The last two terms are opposite in sign, thus their effects on the naturalized flows offset each other. The question thus arises: how large are these gains and losses compared with the gauged flows? If the gains and losses are small in comparison to the gauged flows or if there is much uncertainty associated with either the gains or losses, then it could be argued that the gauged flows provide the best source of information on inflows to Upper Klamath Lake, independent of the coverage and quality of the climate records.

To examine this question, estimates of the annual synthetic natural flow of the Williamson River were taken from Table B-1 of the NFS, and compared with the annual gauged flow, as reported by the USGS. Figure 4-13 shows the time series of the gauged annual flow plotted as a percentage of the estimated annual flow. This plot illustrates an important point about the model's formulation and performance: The ratio of gauged flows to naturalized flows is always less than one, which is to be expected if the modeled losses associated with consumptive use (irrigation) always exceed the modeled gains associated with return flows and decreases in evapotrans-

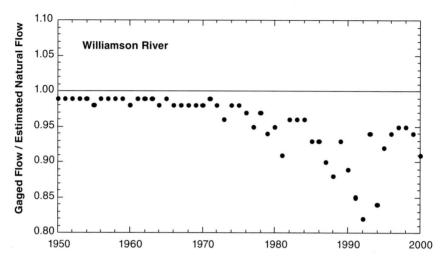

FIGURE 4-13 Measured (gauged) flow plotted as a percentage of the estimated natural flow. Values based on information listed in Table B-1, USBR (2005) and USGS (2007a).

piration from reclaimed marshlands. However, it is also evident that the importance of these losses and gains has changed over time. Prior to about 1980, the modeled losses and gains represent no more than about 5% of the gauged flow of the Williamson River, and since then, the modeled losses and gains have varied between about 10% and 20% of the gauged flow. The uncertainties in modeled losses and gains were not assessed in the NFS, but these uncertainties are likely to be greater than 5%. If that is the case, the trends shown in Figure 4-12 suggest that the NFS model may not be capturing the hydrologic effects of irrigation and reclamation upstream of Upper Klamath Lake until about 1980.

Quality Assurance and Quality Control

The NFS followed a truncated path of the QA/QC process and did not adequately address the issues of planning and conceptualization of the study modeling. The consequences of this apparent shortcoming manifest themselves most prominently through the lack of an overall strategy on how to integrate various modeling efforts, and how to ensure that the basin-wide objectives can be successfully addressed. The work conducted by different agencies and modeling teams involved in the NFS are disconnected, resulting in the development of a model that is inadequate for the task.

Normally, QA/QC involves five phases or steps (NRC 1990). They are

1. Modeling plan for the study, including the problem and its context, study objectives, data availability, modeling needs, performance criteria, and integration.
2. Conceptualization and definition of data needs, including definition of system boundaries, data processing, model structure and parameters, modeling assumptions, and assessment of the model's soundness.
3. Model construction, including development, testing, calibration criteria, and documentation.
4. Calibration and testing of the model, including data collection, analysis, and quality control; calibration strategy; selection of methods; sensitivity and uncertainty analyses; and documentation.
5. Simulations and evaluation, including analysis of results, interpretation, studies of error propagation, reporting, and documentation.

The NFS used a water-budget modeling framework supported by a variety of empirical or semi-empirical models. The study presented the scientific bases underlying the aforementioned empirical and semi-empirical models in a fragmented fashion. To perform water-budget calculations, the USBR compiled an extensive database of stream flow and climatological records. There are no documents demonstrating that the USBR conducted any formal, comprehensive data validation or verification to ensure that the quantity and quality of collected data are adequate for the study. Preceding sections illustrated examples of the lack of such a process, including unverified estimates of groundwater inflows, and unaccounted diversion losses and return flows.

Where the empirical data were unavailable or inadequate, the USBR used expert judgment to provide necessary inputs. Study participants did not assess the reasonableness of these inputs and did not formally and precisely characterize these judgments. USBR Attachment A (2005, p. 5) states that "certain adjustments were made to the marsh ET [evapotranspiration] and crop ET values," and it (p. 12) indicates that "these values compare relatively well." The NFS did not state how it determined reasonableness and appropriateness, and the study provided no formal or informal measures to convey how these values and adjustments were established.

The USBR (2005) study also is deficient in terms of verification and sensitivity analyses. Once the data have been collected, analyzed, and interpreted, some mechanism is needed to ensure that the quality of data-related actions is acceptable for their intended end use. Methods for QA/QC vary by discipline, but there are commonly recognized conventions to ensure high-quality data and scientific output (NRC 2006). Verification and sensitivity analyses commonly characterize, document, and report a model's accuracy. These steps, as well as the others listed in the description of Phases 3 to 5 of the QA/QC process described above, are not present in the documentation of the NFS report.

Uncertainty Assessment and Error Propagation

The NFS (USBR 2005) selected a model based on the water-budget approach to assess the agricultural depletions and alterations to the natural flows of the upper Klamath River. The budget accounted for groundwater flow, natural inflows from the streams above Upper Klamath Lake, predevelopment evaporation losses of Upper Klamath Lake, and predevelopment evapotranspiration losses from marshes surrounding it. The NFS conducted a water-budget assessment, with an Excel spreadsheet, which resulted in monthly average outflows from Upper Klamath Lake at Link River. Corresponding flows at Keno were developed using regression analysis. This section assesses how well the NFS evaluated uncertainty in the water-budget model.

The NFS report should have presented a comprehensive treatment of the method and analytical procedures accounting for uncertainties in the models, parameters, and data. Unfortunately, Chapter 4, *Model Verification, Sensitivity and Uncertainty Analysis*, provides very little information on this subject. There was no attempt to assess the impact of structural weaknesses of the model caused by exclusion of significant factors affecting watershed runoff (for example, clear-cutting in timbered areas, land clearing for pasture and ranching, suppression of fire in forested areas, channeling and diking for flood control and land reclamation, and roadway encroachments). Uncertainty information about key components of the model accounting for evaporation from Upper Klamath Lake and evapotranspiration losses quotes numbers "about 25%" and "about 20 to 25%" respectively, as "range of uncertainty" or "error" in computed numbers. The study provided no explanation or additional documentation explaining how it obtained these numbers.

Similarly, the section of NFS's Chapter 4 discussing stream gaging errors (the only data uncertainty mentioned in the report) provides a very informal estimate of 5% to 10% for the stream flow records of the Sprague and Williamson rivers. The NFS made no attempt to assess uncertainty caused by ungauged watersheds. The section on model sensitivity speculates that "evaporation, evapotranspiration and gaging errors are likely to have the most significant influence in the water budget." There is, however, no information on whether any kind of sensitivity analysis was conducted, and if so, its outcome.

Finally, since the report does not provide any information on overall uncertainty of the modeled stream flow series, it seems that there was no attempt to propagate uncertainties and errors through the model.

In summary, the NFS should inform the users of the report about the ways that errors and uncertainties were, and were not, addressed in the study. This information should be followed by detailed analysis and discussion of the potential impact of errors and uncertainties on the results of the study.

Information Gaps

There are a number of information gaps, that if filled, would greatly enhance the NFS. Improvements in the study, its model, and its output would result by addressing the following shortcomings:

Marsh-Water Limiting Functions. Evaporation from the marsh areas surrounding Upper Klamath Lake is computed using a marsh-water limiting function to account for the varying depth of groundwater table below the surface elevation of the marsh. In the absence of real data, the NFS tested linear, parabolic, and other evapotranspiration extinction functions and chose the linear marsh-water limiting function without explanation for this choice. The function apparently can have significant influences on the outcome of the NFS, as illustrated by one sensitivity study reported to the committee (Perry 2006). In this study, when agricultural acreage in basins tributary to Upper Klamath Lake increased, more water flowed out of the lake. There are two opposing interpretations of this result. The first is that the result simply is wrong, that when agriculture increases, less water should be discharged from the upper lake system because of heavy agricultural consumption. The second and opposite view is that substantial evapotranspiration losses occur from marsh surfaces, so when agriculture is introduced, less water may be consumed due to lower lake levels, and the surface flows from the lake increase. While neither the NFS nor the committee has been able to fully investigate this question, it is possible that the choice of a different marsh-water limiting function would have produced different results. The matter is important enough that it deserves further investigation and explanation.

Groundwater. A more effective NFS requires quantitative information on the upper Klamath basin's groundwater reservoirs, including potentiometric surfaces for all aquifers, hydrogeologic properties, pumping rates, and surface-water-groundwater interactions. The ongoing USGS-Oregon Water Resources Department groundwater study, slated to be completed in late 2007 or 2008, will be a major advancement in addressing the shortcomings in groundwater data.

Forest Cover.   The NFS requires data describing the spatial and temporal changes in forest cover to estimate their hydrologic effects. Aerial photos and satellite imagery, as well as other map and documentary records, provide the sources for such data.

Lower Klamath Lake–Link River.   The regression-equation treatment of the interactions between the Link River and Lower Klamath Lake is inadequate. Additional information (for example, channel geometry, channel slope, roughness coefficients) must be developed such that a rigorous, hydraulic approach can be used.

Hydroclimatologic Change.   The USBR should examine the available information on the hydroclimatologic changes since the beginning of the model simulations to ascertain if these changes could significantly influence the natural flow regime.

Temporal Resolution of Stream Flow.   Monthly stream flow estimates are inadequate to guide decisions about instream flow management. To be useful for instream flow assessment, the NFS model needs to reproduce the entire annual discharge hydrograph with daily resolution. Although the USBR did not intend for the NFS to be used for such a purpose, the study in fact came to be used for this purpose, and it is likely that it will be used again for such purposes.

The appropriate time scale for outputs from the NFS is daily, because daily values of temperature, dissolved oxygen, and discharge are critical to fish habitat. For example, if very low flows persist for 3 or 4 days, the result may be extensive stress on the fish population, especially if those low flows are coincidental with low dissolved oxygen and high temperatures. If the NFS provides only monthly data, the extreme low conditions become lost in the monthly average. The system manager needs to know the daily predicted flows to avoid such undesired conditions, and if only monthly values are available, the manager will be unaware of their likelihood.

Creators of the IFS recognized the value of daily flows as input for their model, based on their importance in modeling extreme events, and requested such values. The USBR was unable to provide daily data because of cost and time required, so that the resulting outputs of the IFS are also constrained by the fact that they deal only with monthly averages.

ALTERNATIVE APPROACHES AND FOLLOW-UP

The approach taken by the USBR to conduct the NFS was but one of many possibilities. Although data availability, funds, and time are limiting factors when choosing an implementation method for a task as complicated as the NFS, the committee would be remiss if it did not suggest alternative approaches or follow-up tasks, in terms of both the overall modeling approach and techniques for estimating the various hydrologic fluxes.

Water-Budget Model

The water-budget approach of adjusting gauged flows to account for evapotranspiration losses and gains is susceptible to significant errors and may introduce unrealistic temporal trends in the naturalized flows. The water-budget model results have not been verified adequately to accept them as reasonable estimates of flows under natural conditions. Moreover, the monthly time step used in the model is inconsistent with the daily data that are ideally required for the development of instream flows at Iron Gate Dam and at downstream locations.

An alternative to the water-budget approach is the development and use of a calibrated rainfall-runoff model for the upper Klamath basin to estimate stream flows under natural conditions. The computation of the naturalized inflows to Upper Klamath Lake using this procedure requires the following general steps:

• Develop and calibrate a precipitation-runoff model for the tributaries of the Upper Klamath Lake using the extensive data that are available for the more recent period of record. This model could include a daily time step and incorporate a groundwater component to adequately simulate the interannual effects of groundwater recharge on stream flows. Hay et al. (2005) developed such a model for the Sprague River basin, a part of the watershed for Upper Klamath Lake. The USBR itself is developing a physically based model that treats not only surface and subsurface flow but also water quality and sedimentation (Matanga et al. 2004).

• Estimate the static input data representative of the predevelopment conditions of tributaries of the Upper Klamath Lake. For example, the land cover representative of the natural conditions may require estimation to simulate such hydrologic processes as evapotranspiration in the watershed model. This model is a "natural system" model of the upper Klamath basin.

• Verify the natural system model for conditions with earliest available records, those from 1904 to 1912, representing conditions closest to predevelopment state.

• Apply the natural system model to the upper Klamath basin using the most recent precipitation records and the potential evapotranspiration estimates to simulate the naturalized flows.

An approach similar to the above has been used successfully for simulating natural flows of the Everglades in south Florida (Fennema et al. 1994).

In this effort, a Natural System Model (NSM) was developed to simulate the hydrologic response of the pre-drainage Everglades. The NSM does not attempt to simulate the pre-drainage hydrologic data that existed prior to human-influenced changes in south Florida, but rather it uses more recent climatic data to simulate the pre-drainage hydrologic response to current hydrologic input. The use of recent input data, for example, rainfall, potential evapotranspiration, and tidal and inflow boundary conditions, allows for a meaningful comparison between the current managed system and the natural system under identical climatic conditions. Information from the managed system models is used to parameterize the NMS.

Using an approach similar to the NSM used in Florida's Everglades is preferable to the naturalization of historical flows used in the NFS, but the ability to calibrate the managed model used to derive the NSM does not ensure a fully calibrated model under natural conditions. The use of parameters calibrated using managed-system models for simulating flows under unmanaged conditions includes some inherent uncertainties that must be considered when such flows are used in any applications.

Evapotranspiration

The scientific literature provides extensive evidence to suggest that the SCS modified Blaney-Criddle method is one of the least accurate methods for estimating evapotranspiration (see review in the evapotranspiration portion of the methods section above). One justification for selecting this method was the insufficient data for other more accurate methods (for example, the Penman-Monteith approach [Monteith 1965]) for the selected period of record. A remarkably more accurate method that does not require as many data as those required for the Penman-Monteith method is the Food and Agriculture Organization (FAO) modified Blaney-Criddle equation (Cuenca 1989, Jensen et al. 1990). Based on statistical analyses of data from many locations worldwide, the FAO modified the SCS modified Blaney-Criddle formula as follows:

$$ET_r = A + B[P(0.46T + 8.13)], \qquad (4\text{-}3)$$

where ET_r is the reference crop (grass) evapotranspiration (mm/day), P is the percent of annual sunshine during month on a daily basis, T is the mean temperature in degrees C, and A and B are climatic calibration coefficients (Cuenca 1989). The coefficients A and B are functions of minimum relative humidity, RH_{min}, the ratio of actual to maximum possible sunshine hours, n/N, and daytime wind velocity, U_{day}. Although the inclusion of these parameters to estimate A and B appear to require a great deal of additional data, Cuenca (1989) suggested that approximate values in the ranges of low, medium, and high can be applied. These approximate ranges of RH_{min} and U_{day} are available from detailed measurements at AgriMet stations (USBR 2007a).

Recent regional reanalysis of historical climatic data through the joint effort of National Center for Environmental Prediction (NCEP) and the National Center for Atmospheric Research (NCAR) has made unprecedented meteorological data available, allowing the use of more accurate combination methods for estimation of evapotranspiration. For example, the North America Regional Reanalysis (NARR) data set (NCEP 2007) provides all the data necessary to apply the Penman-Monteith equation for estimating reference crop evapotranspiration. Another data set has been developed by using the NOAA Land Data Assimilation System (NCEP 2006), which is a 51-year (1948-1998) set of hourly land-surface meteorological data used to execute the NOAA land-surface model, all on the 1/8th degree (about 12 km) grid of the North American Land Data Assimilation System (NLDAS). In this model, the surface data include air temperature, humidity, surface pressure, wind speed, and surface downward short-wave and long-wave radiation, all derived from the National Center for Environmental Prediction–National Center for Atmospheric Research (NCEP-NCAR) Global Reanalysis.

To show how these data could be used for evapotranspiration estimation, the NARR data downloaded for a nine-point grid near Upper Klamath Lake (Figure 4-14) demonstrate how these data could support evapotranspiration estimation.

Figure 4-15 is a box and whisker plot showing how the reference evapotranspiration can be calculated for grass using the Penman-Monteith equation for one of the points shown in Figure 4-14. The daily reference evapotranspiration values computed using climatic data available from regional reanalysis project can easily be used for hydrologic modeling in the Klamath basin. However, when using regionalized data, some local calibration of reference evapotranspiration may be necessary to account for localized effects within the basin.

Agricultural Development

Perry (2006) reported that during the sensitivity analysis of the NFS, the model produced an increase in flow at the Keno gauge when the consumptive use in the upper Klamath basin increases. In other words, *more water used by agriculture means more water going into the Klamath River.* Perry posited that this phenomenon had something to do with "water limiting" conditions for evapotranspiration in Upper Klamath Lake, but offered no further explanation. Although such a result is not impossible, it is at least counterintuitive, and it should be explored and explained completely,

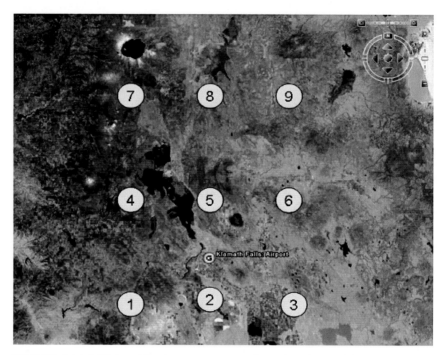

FIGURE 4-14 NARR grid points near Upper Klamath Lake used for downloading climatic data for evapotranspiration estimation.

because it could call into question the underpinnings of the entire model and lead to the unconfirmed conclusion that additional agricultural development would actually increase upper Klamath basin flows.

Groundwater Inflows to Upper Klamath Lake

The NFS estimated inflow to Upper Klamath Lake as the residual in the water-budget equation. This approach, while simple, lumps all the errors into the residual. The USBR did not attempt to verify that its estimates were reasonable, which it could have done by calculating inflows to the lake using a Darcy's Law approach and the existing potentiometric data. Gannett (2007) also presented some preliminary quantitative information on the groundwater flow in the upper Klamath basin.

Changes in Land Cover

The study made no attempt to consider the effects of changing land cover in the upper Klamath basin, especially the loss of forest land and its

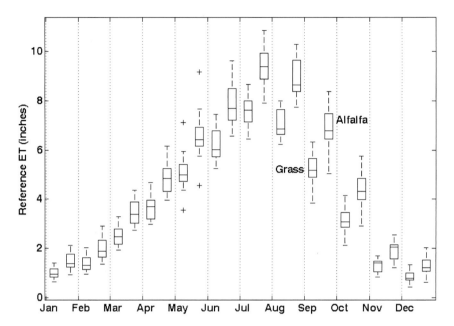

FIGURE 4-15 Box and whisker plots of monthly evapotranspiration (summed from daily data) estimated by using the NARR meteorological data for grid point number 5 (Figure 4-14) and the Penman-Monteith equation.

conversion to other non-forest uses. These changes would presumably affect the infiltration and runoff in the upper-basin watersheds, which could in turn affect the stream flow. The development of a detailed land-cover history of the upper basin watershed using readily available aerial photography supplemented with satellite imagery would enhance the reliability of the resulted estimated flows.

Interactions with Lower Klamath Lake

Perry (2006) indicated that simulation of Lower Klamath Lake needs to be revisited. The NFS concluded that

> Several attempts were made to model this complex system using a digital representation in Excel of the lake/river physical interaction. Because of the complexity of the hydraulics, and the need for detailed data, this modeling is not possible at this time. A correlation approach was used to estimate Keno flows at the outfall of the LKL [Lower Klamath Lake] system. Measured data for the period of October 1904 through September 1918 for the Link River and Keno gauges were used (USBR 2005, Attachment F, pp. 15-16).

Regression modeling is inadequate to simulate the complex interactions between the river and Lower Klamath Lake, so this task will likely require a complex modeling effort.

THE NATURAL FLOW MODEL: CONSEQUENCES

The stated purpose of the NFS was to "estimate the effects of agricultural development on natural flows in the Upper Klamath River Basin" (USBR 2005, Preface). Essentially, the USBR's study provided flow estimates that would be observed if *there was no agricultural development in the upper Klamath basin*. To quote further,

> The approach was to evaluate the changes of agriculture from predevelopment depletions, estimate the effects of these changes, and restore the water budget to natural conditions by reversing the effects of agricultural development. Records used in this empirical assessment were derived from both stream gaging flow histories and from climatological records for stations within and adjacent to the study area. (USBR 2005, p. ix)

Also, due to the effects of the flows required by the NMFS biological opinions, government officials and others saw the NFS as a means to compare pre-development flows with those required by the biological opinions and effect comparisons between the two:

> It seems to be a common belief that the NFS was produced to counter the flow recommendations of the Hardy Phase II report and/or the NOAA biological opinion. As stated previously, the NFS was originally developed as a means of comparing biological opinion flows to natural conditions because that information was requested by federal officials and others. (J. Hicks, USBR Klamath Project Area Office, personal communication, 2007)

Hicks continued with a discussion of the connections among the NFS, ESA considerations, and operation of the irrigation project:

> Members of Dr. Hardy's technical team asked if Reclamation intended to use the NFS to define the maximum releases to the river from the Project with the goal being to mimic the natural hydrograph. An understanding of ESA consultation procedures illustrates that this was and is not the case. In ESA consultation, a required step is to describe the baseline conditions. The baseline is defined in the implementing regulations as current conditions and includes past actions and modifications to the ecosystem. The proposed action, here, continued operations of the Klamath Project, is then compared to the baseline to determine the affect to the species. Any negative effect of the action on listed species is in addition to the conditions described in the baseline. The natural flows, prior to agricultural develop-

ment, are not relevant to the baseline or to the effect Project operations will have on the listed species under the current conditions. The analysis of Project effects is not a comparison of natural conditions versus with-Project conditions, it is a comparison of baseline to baseline plus Project affects and what the consequences are for the continued existence of the listed species. (J. Hicks, USBR Klamath Project Area Office, personal communication, 2007)

It is premature to speculate upon the future uses of the NFS model for future management or policy decisions. The USBR invested heavily in the development of the model, so it would seem unlikely that it will be permanently retired:

> As for other potential decisions, the NFS model is a tool, and like any tool, you must first determine if it is appropriate for use in the particular management or policy decision at hand. It is not prudent to speculate what future management or policy decisions may arise and if the NFS would inform them or not, however, it will be evaluated on a case by case basis to determine when and if it applies to specific decisions and only be used when appropriate. (J. Hicks, USBR Klamath Project Area Office, personal communication, 2007)

The NFS model has already been applied to situations beyond its intended use. It is likely to see further use in studies related to the Klamath Project and listed species so that modifications and enhancements to the model should be made with these potential applications in mind.

CONCLUSIONS AND RECOMMENDATIONS

In conducting the NFS, the USBR faced a daunting task: unravel the complicated natural and artificial "plumbing" of the upper Klamath basin, and from that knowledge estimate the degree to which flows of the Klamath River at Keno Dam would approximate those prior to agricultural development. In addition to the intellectual challenges, the NFS faced time, money, and personnel constraints.

Additionally, the USBR conducted the study in a highly charged atmosphere in which virtually every person and stakeholder group in the Klamath basin, as well as many outside the basin, had an enormous stake. Almost from the beginning, many misunderstood the initial *raison d'être* for the study. Its original purpose did not include providing input for the model under development by the Instream Flow Phase II Study. However, the NFS eventually served this purpose, because little else was available and the IFS also was under tight deadlines. Some groups wanted the USBR to extend its NFS to the entire Klamath basin, and expressed disappointment when that proved unfeasible. The USBR was able to provide the IFS group with "quickly estimated unimpaired flows" for the major creeks as well as the Scott, Salmon, and Trinity rivers.

There are some notable features of the NFS model:

• The conceptual model used for the development of the water budget is thorough and includes attention to detail.
• The seasonality of the simulated natural system flows is adequate at the monthly scale.
• The model is a good representation of a complicated hydrologic system with pronounced anthropogenic modifications, given the time, personnel, and cost constraints and the contentious atmosphere.
• The model captures decadal variations in precipitation and runoff that occur independently of the modified system.
• The model is constructed on a relatively user-friendly platform (Excel spreadsheet) and its modular construction makes it easy to modify and use.

These and other features of the model lead the committee to conclude that it has some utility in providing a generalized picture of unimpaired (natural) flows in the system and in providing a general sense of minimum flows that should be provided to ensure the safety of the basin's fishes, although not precisely enough to lead to day-to-day management of the system. These topics are discussed in more detail in the final section of the chapter. Specific conclusions and recommendations follow.

Conclusion 4-1. The model generally lacks adequate calibration and testing. Overall, QA/QC were not performed in accordance with available scientific and practical standards.

Recommendation 4-1. If the current modeling approach is retained, calibration, testing, and QA/QC need to be performed adequately.

Conclusion 4-2. Treatment of error, sensitivity, and uncertainty did not receive adequate attention and was not performed at the level required for informed decision making.

Recommendation 4-2. If the current modeling approach is retained, document, report, and illustrate the sensitivity analysis of the water-budget model and develop uncertainty estimates of simulated natural flows.

Conclusion 4-3. Lower Klamath Lake was integrally linked to the mainstem Klamath River. High flows in the Klamath River overflowed into

Lower Klamath Lake, effectively attenuating peak flows before they reached the Keno gauge. During the flood recession, there was some "return flow" from Lower Klamath Lake into the main-stem Klamath River, and during low-flow periods the two were disconnected. This natural arrangement ended with the construction of a railroad grade in the early twentieth century that prevented river overflow into the lake, thereby increasing peak flows in the main stem. Simulation of the river's natural flow should therefore include interactions with the lake.

Recommendation 4-3. If the current modeling approach is retained, simulation of interaction between the Klamath River and Lower Klamath Lake is inadequately represented. The present NFS attempts to capture the effect of these complex river-Lower Klamath Lake interactions using only a simple regression to relate flows at the Link River gauge to flows at the gauge below Iron Gate Dam on a monthly time step. A more rigorous model of natural flows should incorporate a hydraulically based sub-model of the Klamath River-Lower Klamath Lake connection on a daily time step.

Conclusion 4-4. Land cover influences hydrologic processes in the upper Klamath basin, hence influences the flows downstream. The NFS does not adequately account for changes in land use and land cover in the watershed, limiting confidence in the model's ability to simulate accurately downstream flows.

Recommendation 4-4. If the current modeling approach is retained, the NFS model should account for the effects of the change in upper Klamath basin forest cover on Keno gauge flows by extending its analysis to historical aerial photography, satellite imagery, ground photographs, and documentary descriptions.

Conclusion 4-5. The prediction of the NFS that flow in the Klamath River at the Keno gauge will increase if irrigated acreage upstream increases is surprising and is not explained.

Recommendation 4-5. The prediction by the model that an increase in agricultural development in the upper Klamath basin produces an increase in flow at Keno gauge is puzzling. The reasons for this model result should be explored, its implications for the model generally should be considered, and the phenomenon should be investigated further and explained.

Conclusion 4-6. The SCS Blaney-Criddle method for estimation of evapotranspiration, a key component of the NFS, is not the best method available and its use in the NFS likely led to inaccurate results.

Recommendation 4-6. If the current modeling approach is retained, the USBR should consider replacing the SCS modified Blaney-Criddle method with a more accurate method, such as the FAO version of the Blaney-Criddle method.

Conclusion 4-7. The results of the NFS illustrate the complexity of hydro-climatic processes within the upper Klamath River basin and show that the seasonal patterns of runoff and storage are strongly influenced by the collective effects of land use and water-management practices. It is apparent that the model captures the longer-term (decadal scale) variations in precipitation and runoff that occur independently of the modified system. The modeled stream flows reflect conditions for the period 1949-2000, which may or may not be representative of the hydrology of the basin. The limited data available for the first half of the twentieth century suggest that precipitation and runoff in the upper Klamath basin were lower then than they were in the latter half of the twentieth century. Future conditions of precipitation and runoff are unknown, thus the volume of water available for agricultural or municipal use, or instream flows, could be significantly different from the flows developed in the NFS.

Recommendation 4-7. Develop a fully distributed precipitation-runoff-groundwater model for the upper Klamath River basin. This type of model is already under development in parts of the basin (Matanga et al. 2004, Hay et al. 2005).

Conclusion 4-8. The monthly time step used in the water-budget model is inconsistent with the daily data required for the development of instream flows at Iron Gate Dam and at downstream locations. A water-budget model like the one used in the NFS is inappropriate for simulation of naturalized flows on a daily basis. The NFS also was limited to the upper Klamath basin, but the IFS requires naturalized flows for the entire Klamath basin.

Recommendation 4-8. Use an integrated framework for the models and their linkages for the entire Klamath basin. Develop a rigorous, physically based precipitation-runoff-groundwater model for the entire basin that can be calibrated using more recent data. The calibrated model can subsequently be used estimate flows under natural conditions.

MANAGEMENT IMPLICATIONS OF THE NATURAL FLOW STUDY

The basic results of the NFS are sets of simulated mean monthly and mean annual flows at Link River and at Keno in cubic feet per second, along with simulated monthly water surfaces for Upper Klamath Lake, under natural conditions. The monthly flows are also expressed as discharges with exceedance values, so that the results indicate values expected to occur greater than 10% of the time, 20% of the time, and so forth up, to 90% of the time. The importance of these values is that, taken together, they represent a statistical picture of flows expected at Link River and Keno if there had been no upstream agriculture and no control by dams. The USBR intended that these flows serve as a representation of the natural flow of the river, although special efforts were made not to underestimate such flows. As such, they might serve as guidance to managers of the system in that they potentially reflect limits of flow below which managers probably should not go to avoid threatening the existence of the system's fish species, and above which they need not go to protect the species.

The implications of the model investigations that produced the simulated hydrologic features of the basin are mixed. From a positive perspective, the results define monthly "natural" variation that managers might reasonably expect, absent their own activities. The monthly variation depicted by the model represents a simulated picture of the conditions under which the biological community of the river evolved and provides a backdrop for assessing the degree to which the present regulated flow regime departs. The flows also provide a general view of the total amounts of water involved in the river and lake regime, with about 1.4 million acre-ft annually flowing out of the lake on the average.

The NFS reasonably captures the decadal variations in flows in the system that are likely to have occurred in the absence of upper-basin development and the installation of dams. These decadal variations, like the monthly variations, are likely to have been ecological features of the biological community as it evolved in the lower river. These variations imply that in the regulated system, some decadal fluctuation in flows is reasonable, and that a completely unchanging regime imposed by engineering structures would not reflect the natural regime.

However, the internal workings of the model in the NFS include several computational shortcomings that imply limits to its use. For example, sensitivity analyses produce unexpected and unexplained responses to changes in the consumption of water in the upper basin. According to the sensitivity analysis, shared with this committee, the flows out of Upper Klamath Lake increase when the agricultural land use in the upper basin is increased. This is a surprising result, discussed above, and at a minimum, it deserves further study and explanation.

In addition, the method for determining evapotranspiration in the model fails to take advantage of recent advances and readily available data that would sharpen predictions. These issues imply that the natural flow model produces results that probably cannot be used as a precise replication of natural flows and that the individual numbers generated by the study are not firm, irrefutable values.

This committee found inadequacies in basic model-building protocol in the construction and refinement of the natural flow model. The model lacks adequate testing, calibration, error analysis, and sensitivity assessment, and it does not adequately address issues related to uncertainty. Although the simulated natural flows have been compared with a short period of earliest gauged records available, the particular approach used for "naturalizing" managed flows does not allow a formal calibration of the model. These shortcomings imply that managers of the biological resources of the basin may use the results of the model in a general way as a form of guidance for the broad characteristics of the natural flow regime, but they cannot use the exact values produced by the study as a template for developing a flow regime with much confidence. Decision makers do not have the advantage of knowing the degree of potential error in the reported monthly flows, values that lack error envelopes or other expressions of uncertainty.

The model is a general representation, and because its output is in monthly time steps, it is not capable of generating the daily time step needed for a completely effective instream flow model to be used in any ecological model downstream. As described in considerations of the IFS in the following chapter, this limitation has a ripple effect that limits the utility of the instream flow recommendations.

Finally, the current model is severely restricted for two general reasons. First, the basin and its biota have changed so much in the past century that the implications for the fishes of restoring "natural flows" are not clear. In addition to changes in vegetation and in the species composition of the animals in Upper Klamath Lake, Klamath River, and their tributaries, the genetic makeup and abundance of the anadromous fishes of the river have changed as well. As a result, it is by no means certain that restoring a natural hydrologic regime to the basin would lead to the distribution, abundance, and species composition that characterized it before the project was initiated. Second, the model does not treat the tributaries of the Klamath River, and they are and have been an essential part of the environments of the anadromous fishes. Without understanding their ecological and hydrological condition and dynamics, it is not possible to understand the ecological and hydrological condition and dynamics of the river.

A modified version of the NFS model, incorporating suggestions made earlier, could have management utility. It could be used as a template for a model of the present-day system. Such a model could be used to simulate "What if?" scenarios, test certain hypotheses, and demonstrate to stakeholders the implications of assorted management decisions and stakeholder choices. Since the NFS model is built upon a familiar, user-friendly platform (Excel), a modified model might find wide use among stakeholders.

5

Instream Flow Study

Results from the Natural Flow Study (NFS) reviewed in Chapter 4 were provided for the Klamath River basin Instream Flow Study (IFS) conducted by Hardy et al. (2006a). Hardy et al. subsequently produced recommendations for instream flows at the U.S. Geological Survey (USGS) stream gauge below Iron Gate Dam by conducting an elaborate series of investigations and model-building efforts, which are reviewed in this chapter. The following pages address the general technical elements of an IFS, and describe the procedures followed by Hardy et al. and the committee's evaluation of those procedures. The chapter continues with an examination of the implications for implementing the instream flow recommendations, followed by brief comments regarding the larger context of those recommendations. The chapter closes with conclusions and recommendations.

TECHNICAL ELEMENTS OF AN INSTREAM FLOW STUDY

Instream flow is simply the water flowing in a stream channel (Annear et al. 2002). This simple concept belies the difficulty in determining what is the most appropriate instream flow regime when considering competing uses for water, such as irrigation, public supply, recreation, hydropower, and aquatic habitat; and the consequences of flow-level changes across seasons and years. Since natural-resource managers must make defensible decisions that balance competing demands for water, it is critical that appropriate and well-documented methods are used to quantify instream flow needs (Annear et al. 2002).

Instream flow programs involve technical and nontechnical components for developing and negotiating acceptable flows. Within a particular IFS, major technical elements include hydrology and hydraulics, geomorphology and physical processes, aquatic resource biology, and water quality (Annear et al. 2004). Each of these technical elements may involve an independent study, though integration of these elements must occur to address connectivity, scaling, integration, quality assurance and quality control, and model testing. Nontechnical components of an instream flow program include legal, regulatory, and public-participation issues that are unique to a particular study. The following subsections provide overviews of the major elements of an instream flow program (NRC 2005a).

Hydrology and Hydraulics

The "natural flow regime" of a river is the characteristic pattern of flow quantity, timing, rate of change of hydrologic conditions, and variability across time scales of hours, days, seasons, years, and multiple years, all without the influence of human activities (Poff et al. 1997). The natural flow regime in general has four components (not all of these necessarily occur in every river or even in every reach of a river) (NRC 2005a): subsistence flows, base flows, high-flow pulses, and overbank flows (Figure 5-1).

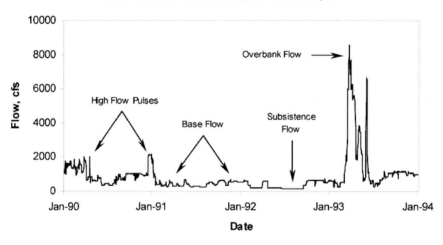

FIGURE 5-1 Illustration depicting flow regime components (subsistence flow, base flow, overbank, high-flow pulse).
SOURCE: Hardy et al. 2006a. Reprinted with permission; copyright 2006, Utah State University.

Subsistence flow is the minimum stream flow needed during critical drought periods to maintain tolerable water-quality conditions and to provide minimal aquatic habitat space for the survival of aquatic organisms. *Base flow* is the "normal" flow condition between storms. Base flow sustains habitat that supports diverse, native aquatic communities. Base flow also maintains the groundwater level that supports riparian vegetation. *High-flow pulses* are short-duration flows confined to the stream channel and occur during or immediately after storms. High-flow pulses flush fine-sediment deposits and waste products from the system, restore normal water quality following prolonged low flows, and provide longitudinal connectivity for species movement. *Overbank flow* is an infrequent, high-flow event that breaches riverbanks. Overbank flows may restructure the channel and floodplain, recharge groundwater tables, deliver nutrients to riparian vegetation, and connect the channel with floodplain habitats that provide additional food and space for aquatic organisms. In contrast to the once popular convention of developing a single, minimum flow or "flat-line" flow, current instream flow science advocates including all flow components in an instream flow recommendation. The many flow components, including maxima, minima, duration, frequency, and timing, are important to managers because various species of interest use habitats that are defined by these flow characteristics, and the best way to gain understanding of desirable attributes of flows is to understand "natural" flows that once supported useful, multi-species habitat.

Table 5-1 demonstrates how the four components of a flow regime affect the major technical elements of an instream flow program (geomorphology and physical processes, biological processes, and water quality). Stream flow has been described as the "master variable" (Poff et al. 1997), because the major technical elements depend on it and because it indirectly controls aspects of the physical, chemical, and biological environment, such as water temperature, hydraulic conditions, habitat, nutrient concentrations, aquatic vegetation, and connectivity.

The goal of a hydrologic evaluation in an IFS is to understand and quantify those processes that affect stream flow quantity. This evaluation includes quantifying the magnitude, frequency, timing, and duration of the four flow-regime components; descriptive aspects of the hydrologic system, such as the location of springs, tributaries, and dams; and the impacts of land and water use on the flow regime (NRC 2004a). The purpose of hydraulic modeling is to define stream flow characteristics (for example, depths and velocities) as a function of discharge and channel and floodplain geometry (NRC 2005a). Results from the hydrologic and hydraulic modeling, along with water-quality and fish-population modeling results (discussed later), facilitate assessing biological, water-quality, and physical processes for managing instream flows.

TABLE 5-1 Flow Regime Components and Their Effect on Physical Processes, Biological Processes, and Water Quality

Flow Regime Component	Geomorphology and Physical Processes	Biological Processes	Water Quality
Subsistence flow	Increase in deposition of fine particulate materials	Aquatic habitat is restricted, vegetation encroachment	Dissolved oxygen decreases, temperature increases, establishing tolerance limits
Base flow		Base hydraulic conditions for aquatic habitat	Near optimal temperatures for physiological processes
High-flow pulse	Flushing of fine sediment, connection to low-level off-channel water bodies, channel maintenance, scouring pools, and uprooting vegetation	Recruitment events for water-propagating species	Increasing levels of bacteria, total suspended solids
Overbank Flow	Floodplain construction and maintenance; connection to off-channel water bodies; bar building, channel migration, and alterations; large woody debris recruitment and transport	High connectivity between aquatic and floodplain systems, yielding biotic exchanges between channel and floodplain and refuge from high in-channel velocities	Increases in total suspended solids and sediment loads

SOURCE: Adapted from NRC 2005a.

One principle of an effective instream flow prescription is to mimic, to the extent possible, the processes characteristic of the natural flow regime (Annear et al. 2002). Meeting this objective requires data for evaluating historical and post-development stream flows, water budgets, and the like. Results from the data evaluation should serve to identify those aspects of the present conditions that must be preserved or enhanced. If there are no suitable stream flow data or if a sufficient period of record is not available, synthetic data generation may be required (Bovee et al. 1998, Wurbs and Sisson 1999). Hydrologic simulation models (for example, HEC-HMS, WMS) use information on watershed characteristics, precipitation, and run-off patterns to synthesize or extend stream flow records or create synthetic ones. If stream flow data are available from gauges in the region, runoff patterns for the watershed of interest can be synthesized by establishing statistical relationships with similar watersheds.

Poff et al. (1997) pointed out that the natural flow regime of virtually all rivers is inherently variable and this variability is critical to ecosystem functioning and native biodiversity. Year-to-year variation in flow regimes drives important physical and biological processes that periodically reset geomorphic conditions; temperature patterns across seasons; and important biological processes, such as fish-egg maturation, incubation, and growth rates characteristic of good and poor year classes (Trush et al. 2000). Therefore, to ensure sustained biological diversity and a functioning dynamic ecosystem, both inter- and intra-annual flow regimes that attain the critical threshold levels necessary to drive important ecological processes must be maintained or provided through managed flow releases (Annear et al. 2004).

Some aquatic species thrive during high-flow water years, while other species do well during years of drought. Generalist species flourish under wide-ranging flow conditions. High flows route coarse sediments, build bars, erode banks, flush fine sediments, scour vegetation, undercut and topple large woody riparian vegetation, all of which are necessary aspects of dynamic rivers that characterize the coastal salmon-rearing streams of the western United States. Typically, anadromous salmonids have successful year-classes during normal to below-normal water years when flow conditions are relatively steady during the spawning, incubation, and fry-rearing seasons. The most favorable habitat conditions usually develop the year after wet years, which scour pools, recruit large woody debris, flush fine sediments, and build bars.

Geomorphology and Physical Processes

Physical processes form and maintain the shape of the stream channel and floodplain (the strip of land that sometimes borders a stream channel and that is normally inundated during seasonal floods, Bridge 2003). The form of a river channel results from interactions among discharge, sediment supply, sediment size, channel width, depth, velocity, slope, and roughness of channel materials (Leopold et al. 1964, Knighton 1998). Sediment transport and deposition also shape the floodplain and riparian zone. Stream channels react to changes in sediment dynamics and either degrade or aggrade along the longitudinal gradient in response to sediment load. Channel form provides the physical structure for habitat for aquatic organisms. Human modifications, such as channelization, bank fortification, and reduction of coarse-sediment load due to dam installation influence the channel form and resulting habitat. Instream flow technical evaluations of physical processes are useful in documenting changes in channel structure, aquatic habitat composition, riparian vegetation, and other effects of the flow dynamics in river systems.

Biology

Instream flow studies historically were single-species oriented, and focused on periodicity of instream life history, physical-habitat suitability, water-quality tolerances, and temperature effects on reproduction, growth, and physiology. This approach led to flow prescriptions that ignored the needs of other fishes and other organisms such as benthic macroinvertebrates, aquatic macrophytes, and riparian species that are dependent on riverine processes (Annear et al. 2004). Given the connectivity among the elements of riverine systems (Vannote et al. 1980), a restricted species-habitat approach could have harmful effects within the very systems the instream flow advocate is trying to protect. Later advances in addressing fish-community needs led to the development of a guild approach to determine the physical-habitat needs of groups of fishes or invertebrates that use similar habitat types such as slow-moving pools or high-gradient riffles (Leonard and Orth 1988, Lobb and Orth 1991).

Recently, instream flow prescriptions have addressed more-comprehensive aspects of riverine system ecology by attempting to meet the flow requirements of the entire aquatic community and the associated terrestrial (that is, riparian) community (Moyle et al. 1998; Annear et al. 2004; NRC 2005a). These holistic efforts are often hampered by limited data on habitat requirements, biology, and life history of aquatic organisms including federally listed endangered species and nongame fishes (Myrick and Cech 2000; Moyle 2002), and by the limited, but growing understanding of the intricate connections within the biological community of a river ecosystem.

Hydraulic habitat (that is, flow depth, velocity, substrate, and instream cover components of a stream) is a key component of any instream flow prescription, but providing hydraulic habitat alone will not guarantee any particular state of the aquatic ecosystem (Annear et al. 2004, NRC 2005a). Appropriate physical habitat is necessary, but not sufficient on its own. The dynamic effects of varying levels of hydraulic habitat on biological processes, including competition, bioenergetics, predation, disease, and the recruitment of juveniles into the population, must be considered (Bartholow et al. 1993, Annear et al. 2004, NRC 2005a). Ecological and biological processes occur over variable scales of time and space, so an instream flow prescription should provide an appropriate level of spatial and temporal variability, to preserve the complexity of these processes.

Water Quality

Historically, the water-quality component of instream flow prescriptions was based on modeling efforts to ensure that water-quality standards (for example, dissolved oxygen, temperature, and nutrient and contaminant concentrations) were not violated. This focus on water-quality standards alone, including the use of 7Q10 to establish minimum flows, often resulted in minimum flow prescriptions that ignored the physiological needs of fishes and other organisms, such as benthic macroinvertebrates, aquatic macrophytes, and riparian species, which also are dependent on riverine processes (Annear et al. 2004). Comprehensive methods address seasonal requirements for successful spawning, incubation and growth of important species of fish, or sometimes a group of fishes such as salmonids (Bovee et al. 1998, NRC 2005a).

Issues of spatial and temporal scale and inter- and intra-annual variability also are relevant to the water-quality component of an instream flow prescription. The influence of water quality on ecological and biological processes occurs at various scales, and responds differently to the flow regimes of wet and dry years. For example, degree-day accumulation determines the timing and location of spawning, egg maturation, and the duration of incubation, fry emergence, and growth for most riverine fishes.

The flows specified by an instream flow prescription exert a significant influence on the temperature regime and water quality within the system, influencing directly or indirectly the biological, physical, and chemical characteristics (Annear et al. 2004, NRC 2005a). Ignoring any of these effects is fraught with risk, because the combined effects of the various components of water quality will influence the presence, abundance, and distribution of biota. The primary water-quality parameters considered by instream flow studies are sediment, total dissolved solids, dissolved oxygen, water temperature, pH, contaminants, and nutrients, but analyses focus on water temperature, contaminants, and sediment (Annear et al. 2004, NRC 2005a). Brett (1964) described water temperature as the "ecological master variable" because it affects all aspects of the biology of aquatic organisms including reproduction (Stonecypher et al. 1994), growth (Jobling 1997), susceptibility to disease (Antonio and Hedrick 1995), and migratory ability (Lee et al. 2003). Clearly, it is an important driver of riverine ecosystems. Any alteration of the flow regime also alters the temperature regime. As with the hydraulic component, the temperature is a necessary but not sufficient component of the habitat for successful instream flow prescriptions.

A broad range of pollutants are present in riverine systems; their effects range from sublethal changes in physiological performance (Beaumont et al. 1995), to disruption of the endocrine system with subsequent changes in reproductive success (Jobling et al. 1998), to changes in community composition (Hickey and Golding 2002), to acute toxic effects (Hamilton and Buhl 1990). The effects of suspended solids and sediments on stream-dwelling organisms are well documented (see Waters [1995] for a comprehensive review) and include loss of spawning habitat (Burns 1970), increased physiological stress (Servizi and Martens 1992), and direct mortality of fish (Servizi and Martens 1991) and other aquatic organisms. Sediments are a major pollutant in U.S. waters (EPA 1990) and should be an important consideration in instream flow studies. Predicted changes in the nutrient status of streams should also be considered in instream flow prescriptions. Streams of particular concern include those that are effluent-dominated or that flow through nutrient-rich landscapes. Nutrient enrichment (or impoverishment) can change the flow of energy in a river ecosystem (sometimes with positive effects; see Bilby et al. 1998), and can alter conditions sufficiently to shift the balance among competing species. Nutrient enrichment may also lead to changes in other water-quality parameters, including dissolved oxygen levels that will have additional negative impacts on river ecosystems.

The development of an instream flow prescription will benefit from prior knowledge of water-quality conditions, but this information is not always available. Monitoring water quality in even river system of modest size is challenging because of the number of physical and chemical parameters that should be tracked, and the heterogeneous nature of those parameters, particularly when spatial and temporal variability, and the heterogeneous legal and jurisdictional framework are included (Woodling 1994). Nevertheless, because of the significant effects flows can have on water quality, instream flow prescriptions should attempt to address the key aspects of changes in water temperature, dissolved oxygen, nutrient concentrations, sediment loads, and contaminants.

Connectivity

Historically, low-flow fish passage for migratory species (for example, anadromous salmon and trout) was frequently the only consideration of biological connectivity when developing instream flow prescriptions. The temptation is strong to simplify riverine ecosystems as unidirectional, two- or three-dimensional systems. For some exploratory modeling exercises, this degree of simplification may be appropriate. However, such simplifications do not incorporate connectivity in lateral, longitudinal, vertical, and temporal dimensions in a river system (Ward 1989). Connectivity is "the flow, exchange, and pathways that move organisms, energy, and matter through these [river] systems" (Annear et al. 2004). The presence of physical, chemical, or biological barriers degrades the connectivity and functioning of rivers. Common connections include nutrient and energy flow from headwaters to downstream (Vannote et al. 1980) and connections between surface flow and groundwater. Connections are severed by dams and culverts (Helfrich et al. 1999, Schlosser and Angermeier 1995), and by changes in flow regimes or water quality (Cherry et al. 1978). These connections should be considered in an instream flow evaluation to ensure that connectivity is improved or not further degraded (for example, by prescribed releases of poor-quality water that effectively impedes upstream movements of fish and invertebrates).

Spatial and temporal scale and intra- and inter-annual variability need to be considered in an assessment of connectivity (Kondolf et al. 2006). Biological connectivity, such as unimpeded upstream passage and access to the floodplain for fishes, does not have to be permanent, but needs to exist during critical phases of fish life history, such as spawning migrations or juvenile rearing (for example, Maslin et al. 1997). Furthermore, the natural flow regime may have included periods of intermittent flow that provided or created off-channel habitat for locally adapted species (Labbe and Fausch 2000); these periods of intermittent flow (and the challenging physical conditions that result) may also afford native species an advantage over invasive nonnative fishes.

Scaling

Physical, chemical, and biological processes affect stream ecosystems at different spatial and temporal scales (a more general discussion of scaling is in Chapter 3). The importance of spatial and temporal scales and the study of flow, temperature, and habitat time series during instream flow studies has been recognized for more than three decades, but seldom has been adequately incorporated into instream flow prescriptions (Bartholow and Waddle 1986, Milhous et al. 1990). Spatial scaling issues, such as specifying what length of a river must be studied, how study reaches are selected, and how data from study areas are extrapolated to unstudied areas, remain a major research focus for instream flow science, and effective methods for reconciling different scales are not well documented.

Spatial Scaling

Habitat-suitability modeling integrates results from hydraulic simulations and fish-distribution modeling. A basic assumption in habitat-suitability modeling is that the hydraulic variables (flow depth and velocity) and structural elements (bathymetry, substrate size, and cover distribution) are uniform within each of the simulated small areas within the gridded design for sampling stream reach. These simulated small grids are often referred to a cells or meshes by the modelers. Consequently, the spatial resolution is a critical element and must be compatible between the two modeling efforts. Often, the cell size determined from field measurements and used for fish-habitat simulations is selected to facilitate hydraulic modeling and calibration. This is especially true for one-dimensional hydraulic modeling when some analysts survey a single transect over individual mesohabitat types (for example, riffle, pool, and run). Failure to consider the spatial scale of fish use and the field-collection techniques used in developing habitat-suitability criteria can result in widely spaced transects (used with one-dimensional hydraulic modeling) yielding rectangular cells many meters in length. Such excessively large cells violate the assumption of cell homogeneity. For fish, an appropriate cell size often is approximately a few square meters (depending the size of the fish and their degree of territoriality and movement), which requires that transects be placed close together (Bovee 1982, Bovee et al. 1998). Two-dimensional hydraulic modeling can more easily overcome this limitation when detailed bathymetry of the stream channel is known. The only assurance that model cell size is compatible is by comparing habitat-model output (habitat suitability by cell) with independent field observations of fish distribution. Statistical tests for acceptance or rejection of model output should be made before using these types of models as input to habitat time series and fish-population modeling. In addition, as described in Chapter 3, the scale for habitat-suitability criteria should be the same as that for hydraulic modeling.

Temporal Scaling

Time-series modeling of habitat and water quality facilitate the evaluation of the suitability of environmental conditions for supporting the completion of fish life stages and yield input to fish-population modeling. These models require weekly time steps at a minimum and ideally would use daily time steps. When rapid flow fluctuations may occur, as under hydropower peaking operations, hourly time steps are necessary.

Integration

Integration is the process of combining different technical studies into flow recommendations (NRC 2005a) at specified points within the river system. Integration requires that evaluations of hydrologic, hydraulic, geomorphologic, biological, and physical processes and water quality be compatible and at commensurate or complementary spatial and temporal scales. Integration of study results and model linkages is an important, complicated step, and although integration methods are being generated empirically, they are not well documented. There are few widely recognized general procedures for integration, and thus methods must be evaluated without the benefit of direct, observable, or historical evidence. The Instream Flow Incremental Methodology (IFIM) (Stalnaker et al. 1995) promotes integration by providing guidance for combining microhabitat (hydraulic habitat) with macrohabitat conditions (water quality) over longer stream lengths through a river system (Bartholow and Waddle 1986,

Bartholow 1989, Bovee et al. 1998). The USGS SALMOD model integrates hydrologic processes, hydraulic-habitat carrying capacity, and degree-days as the principal drivers of a salmon population-dynamics model (Bartholow et al. 1993). The integration process is not achieved by simply prescribing flows that mimic the shape of the natural flow regime. Appropriate integration methods link instream flow recommendations to desirable outcomes using scientifically defensible methods. An example of such integration and linkage can be found in the Trinity River Flow Evaluation report (US-FWS/HVT 1999), considered a founding document for the Trinity River Restoration Program.

Quality Assurance and Quality Control

As described in Chapter 2, quality assurance and quality control (QA/QC) for ecosystem modeling requires a modeling plan, adherence to procedures, documentation, and model testing. Given the importance of evaluating alternative management schemes, rigorous QA/QC protocols, including peer review, should be established, documented, and enforced. Also, if legal challenges are likely, consideration should be given to establishing and documenting model and data versions, how data were processed and modified when necessary, and the custody of information during the modeling process.

Model Testing

Model testing refers to procedures used to evaluate the performance of predictive models (Chapter 3). Instream flow studies inevitably utilize a multitude of models including hydraulic (for predicting flow depth and velocity), water quality, water temperature, habitat suitability (for various species and life stages), water routing, and reservoir operation models.

Model testing requires that the model produce testable predictions and that independent data are available for comparing model output to observations. Model test conditions should span the entire range of conditions for which the model will be applied. In the case of hydraulic models, for example, comparison between observed and predicted velocities is often tested for flow within the channel banks, and comparisons are seldom made for overbank flows. Because a comprehensive IFS must examine overbank conditions, hydraulic models should be tested under those conditions. Habitat models rarely are tested against observed fish distributions in a modeled stream reach. Thus, the use of hydraulic-habitat-suitability models has become quite controversial and has been criticized. Testing of model output is essential to establishing that a model is reliable. Box 5-1 provides the committee's recommendations for improving.

BOX 5-1
Critical Methodological Issues for the Analysis
of Habitat-Instream Flow Relationships

Below the committee describes four important issues in the analysis of habitat and instream flow. They include some methods that could improve the IFS's modeling by incorporating methods more broadly used by ecological modelers. The key citations on habitat-selection modeling are Manly et al. (2002) and the examples of applications collected in Scott et al. (2002).

1. It is important to select a single, biologically appropriate spatial resolution and use it for all parts of the model. Ecological relations, such as how habitat affects fish, vary with spatial scale; understanding this is a fundamental aspect of ecology (Levin 1992). For example, a relatively fine resolution is needed for small, territorial fish, while larger resolution, perhaps even whole pools or riffles, may be appropriate for large fish that routinely move large distances (Kondolf et al. 2000). A fatal flaw in habitat modeling is the practice of generating "habitat-suitability criteria" (relationships between micro-habitat variables such as depth, velocity, substrate, and cover and the "suitability" of habitat for fish) by measuring habitat at a fish's exact location (a very fine resolution), then applying the criteria to cells several square meters in size. The mismatch in scale between the habitat criteria and the cell size undoubtedly induces significant errors, and it is very difficult to estimate the magnitude of the errors. Instead, a study following the fundamental rule of choosing a single scale would start by selecting a cell size appropriate for the fish species and life stages of interest. Then data for habitat criteria would be obtained by observing the density of fish, and cell-averaged habitat variables, in a number of cells (Manly et al. 2002). The same cell size would then be used in the hydraulic model. Railsback et al. (2003) provided an example in a virtual stream for a trout population. In the IFS, the hydraulic model had cell sizes of 1.7×1.6 m, and the habitat model had 0.6-m cells (Hardy et al. 2006a,b); the simulated hydraulic properties were used "to generate habitat computational meshes at a 0.6-m grid spacing using bi-linear interpolation for bed elevations, depths and velocities" (Hardy et al. 2006b). Thus, the IFS did not use the same cell size for hydraulic and for habitat modeling, and a more-explicit description of its rule for choosing cell size would have been helpful. However, preliminary tests of the model showed that the distribution of Chinook salmon

---

OVERVIEW OF PROCEDURES USED IN
THE INSTREAM FLOW STUDY

Background

An interim Klamath River IFS (Hardy 1999), Phase I, reviewed the historical and current status of anadromous fishes in the lower Klamath River. Phase I also highlighted likely factors leading to the decline of the fisheries and made interim minimum monthly flow recommendations for the mainstem Klamath River from Iron Gate Dam to Scott River. The recommended

fry was consistent with the hypothesized relationships between instream flow and habitat (Hardy et al. 2006b). Expanding the model to cover other life stages and species will require careful application of ecological principles and explicit documentation.

2. Use an index of habitat selection (or "preference") that has a clear biological meaning and is measurable and testable in the field. Manly et al. (2002) recommended that habitat-selection models simply predict the density of animals (number of individuals per square meter) in each habitat cell. This provides testable predictions instead of WUA, which has no clear biological or statistical meaning and cannot be tested in the field. WUA is considered to be an index of "suitability," but it describes only how often fish are observed in that habitat type, not that a habitat type is suitable in any particular way. Railsback et al. (2003) illustrate the use of fish density as an alternative to WUA.

3. Use multivariate, information-theoretic, statistical methods to develop the habitat-selection ("preference") functions. Models used in the IFS are based on "suitability criteria" that treat the effect of each habitat variable as independent, and assume all variables have equal effects (because these criteria are all forced to range from zero to one). Guay et al. (2000) found that using logistic regression instead of PHABSIM methods could double the correlation between modeled habitat preference and observed fish densities. Manly et al. (2002) and others provided several approaches; Guay et al. (2000) and Ahmadi-Nedushan et al. (2006) addressed this issue specifically for PHABSIM-based studies (although they did not deal correctly with the spatial resolution issue). The commercial HyperNiche software (www.digisys.net/~mjm/hyperniche.htm) appears to provide a very powerful and easy-to-use approach to the development of habitat-selection functions.

One of the primary benefits of such techniques for modeling habitat selection is that they provide estimates of uncertainty in the habitat relationships. Given this ability, failing to develop and analyze uncertainty estimates is very hard to justify.

4. Address the fact (documented in this chapter) that habitat-selection relations can vary strongly with environmental and biological variables, such as temperature, turbidity, habitat structure, and fish density. As a consequence, mixing habitat-suitability data from a variety of sites and conditions may produce criteria that represent no site. Habitat-selection studies should not be considered acceptable without clear evidence that the habitat-preference functions are appropriate for the study site.

flows were higher than previously established FERC flow regimes and more closely resembled the shape of the natural flow hydrograph (Hardy 1999). The flow recommendations specified in Phase I were considered to be interim until a more ecologically based flow regime was developed through Phase II of the study (Hardy et al. 2006a), which is reviewed here. The Phase II study sought to recommend a flow regime that would protect the physical, chemical, and biological processes necessary to restore and maintain aquatic resources of the Klamath River.

Specifically, the stated purpose of the Phase II study is

to recommend instream flows on a monthly basis for specific reaches of the main stem Klamath River below Iron Gate Dam by different water-year types. These recommendations specify flow regimes that will provide for the long term protection, enhancement, and recovery of the aquatic resources within the main stem Klamath River in light of the Department of the Interior's trust responsibility to protect tribal rights and resources as well as other statutory responsibilities, such as the Endangered Species Act. The recommendations are made in consideration of all the anadromous species and life stages on a seasonal basis and do not focus on specific target species or life stages (i.e., coho). (Hardy et al. 2006a, p. ii)

The report also states,

the primary objective for Phase II was to develop instream flow recommendations using best available science employing state-of-the-art field data collection and modeling techniques. (Hardy et al. 2006a, p. 49)

The extensive research for Phase II used a collaborative process involving Utah State University and an interagency technical review team. The review team consisted of representatives from the U.S. Fish and Wildlife Service; National Marine Fisheries Service; U.S. Geological Survey; Bureau of Indian Affairs; Bureau of Reclamation; California Department of Fish and Game; Oregon Department of Water Resources; and the Karuk, Hoopa, and Yurok Tribes. This federal, state, and tribal technical team reviewed the Phases I and II reports. Experts, primarily local resource-agency biologists, provided their judgment periodically to make decisions about the adequacy of habitat model results.

Overall, Hardy et al. (2006a) represents an advanced assembly and extension of what is known about the behavior of salmon fry and the suitability of reaches in the main-stem Klamath River for several life stages of Chinook (*Oncorhynchus tshawytscha*) and coho (*O. kisutch*) salmon and steelhead (*O. mykiss*). Although this assembly has consolidated and advanced the state of knowledge and the ability to quantitatively represent the potential of the main stem to support these species and life stages (both locally and overall), it is, like all models, an imperfect representation of reality.

Field-Data Collection

The flow diagram presented in Figure 5-2 presents an overview of how Hardy et al. (2006a) divided the main-stem Klamath River into reaches, identified study sites within each reach, and then collected field data in the reaches and study sites for producing instream flow recommendations. Following the general methods specified in IFIM, Hardy et al. (2006a)

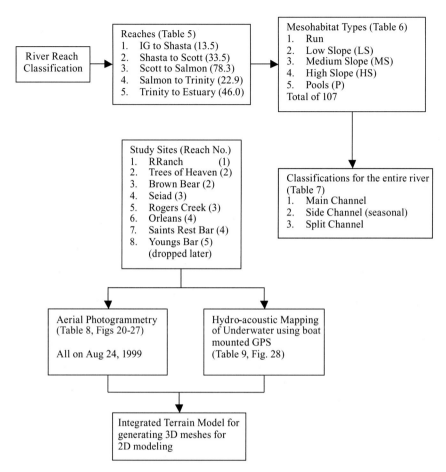

FIGURE 5-2 Overall data flow diagram of field data collection step in the Phase II study of the IFS.
SOURCE: Hardy et al. 2006a. Reprinted with permission; 2006, Utah State University.

delineated the Klamath River below the Iron Gate Dam into five river reaches. Stratification of the river into "homogeneous" study reaches was primarily based on junctions of major tributaries with the lower Klamath River. Figure 5-3 shows a representative reach along the Klamath River, below Iron Gate Dam.

Hardy et al. (2006a) selected eight study sites within the five river reaches. They chose study-site locations that were broadly representative of the channel characteristics within each delineated river reach and, in

FIGURE 5-3 This reach of the Klamath River between Iron Gate Dam and the abandoned settlement of Klamathon is typical of the representative reaches used in the IFS.
SOURCE: Photograph by W.L. Graf, University of South Carolina, July 2006.

some cases, to overlap with existing study sites. Hardy et al. (2006a) also considered data from ground-based habitat mapping in the site-selection process. At the study sites, data were collected intensively to facilitate application of hydrodynamic and habitat models. One site, Young's Bar, was later dropped from the study because efforts to calibrate the hydrodynamic model developed for the site were unsuccessful, so seven study sites were used in the IFS.

Using five categories of mesohabitat type (runs, pools, low slope, medium slope, and high slope), habitat mapping of the entire river below Iron Gate Dam was undertaken by the USFWS, USGS, and Yurok Tribe. Relationships between discharge and available fish habitat were scaled from specific study sites to the reach level by using the habitat-mapping results.

Hardy et al. (2006a) acquired low-elevation high-resolution aerial photography and underwater hydro-acoustic maps. These data were combined into an integrated terrain model that was used for hydrodynamic and habitat modeling. This effort represents state-of-the-art field data collection.

Fish-habitat utilization data were collected at study sites to fulfill two objectives. The first objective was to provide data for development and testing of the conceptual physical habitat models, and validation for the habitat modeling results. The second objective was to obtain sufficient data to develop site-specific suitability criteria for use in the habitat models. Several sampling protocols were used to collect fish-habitat data depending on the species and life stage of interest.

Hydrologic and Hydraulic Considerations

Natural and Impaired Flows

After a brief introduction, Hardy et al. (2006a) assess developments, including channel modifications and dike construction that changed the historic flow regime in the Klamath River and that confound present estimates of historical (1905-1912) mean monthly flows at Keno Dam. Although the historical flow data were derived from gauge records and adjusted for above-normal precipitation during this period, Hardy et al. (2006a) note that their value is questionable because of the impact of developments and the short period of record. Because of these concerns, in the IFS, use of the historical flows is limited to comparisons with (1) impaired flows derived from gauge records and (2) estimated natural flows from the various Natural Flow studies sponsored by the USBR. Results from the comparisons indicate that both the timing and magnitude of flows have changed at the dams because of Klamath Project operations.

The IFS uses mean monthly flow estimates from two natural flow studies conducted by the USBR. The first study used level-pool routing in Upper Klamath Lake based on observed net inflows between 1961 and 2004 with adjustments to account for estimated consumptive uses (PWA 2002; Hardy et al. 2006a, Appendix A). In the consumptive-use study, flows from Klamath Lake were routed to Iron Gate Dam, assuming no Klamath Project demands and using accretions currently used by the USBR in its operations models for the Klamath Project. In addition, the study contained natural flow estimates for Link River and Iron Gate Dam. From the consumptive-use study, Hardy et al. (2006a) used the estimated flows for the Keno gauge. To obtain flows at and below Iron Gate Dam, Hardy et al. (2006a) added estimates of unimpaired accretions to the USBR flow estimates for the Keno gauge.

In the second NFS, the USBR derived natural flow estimates at Keno, Oregon, using a water-budget approach to assess the effects of agricultural development and other alterations to the natural flow (USBR 2005; also see Chapter 4). Again, to obtain flows at and below Iron Gate Dam, Hardy et al. (2006a) added estimates of unimpaired accretions to the USBR natural-flow estimates at Keno in order to derive flow estimates at down-stream locations.

A hydrograph showing mean monthly flows at Iron Gate Dam was developed by Hardy et al. (2006a) and is reproduced in Figure 5-4. The figure shows adjusted historical flow data (1905-1912), natural-flow estimates from the two USBR studies (unimpaired consumptive use and NFS), and gauge record flows at Iron Gate for the period from 1961 to 2000 (Iron Gate Impaired). Hardy et al. (2006a) note that flow estimates derived from the historical data follow the same pattern as the natural flow estimates, albeit with some systematic error. Based on the similarity of the natural-flow estimates from the two USBR studies, Hardy et al. (2006a) conclude that differences in the estimates are due to differences in the methodologies. Hardy et al. (2006a) also note that (1) the level-pool, consumptive use approach generates higher monthly flow estimates than the water budget approach. From a plot showing annual flow duration curves at Iron Gate Dam (Figure 5-5), Hardy et al. (2006a) note that for impaired flow conditions, high flows (discharges that are exceeded less than 10% of the time, called exceedance values of less than 10%) are higher and low flows (discharges with exceedance values greater than 30%) are lower.

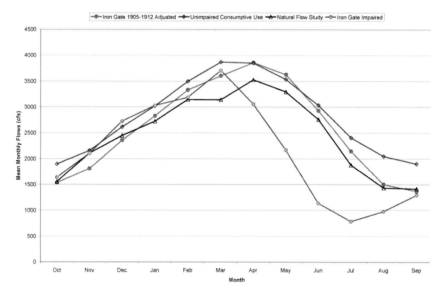

FIGURE 5-4 Estimated mean monthly flows and observed flows (1961-2000) at Iron Gate Dam.
SOURCE: Hardy et al. 2006a, Figure 9. Reprinted with permission; copyright 2006, Utah State University.

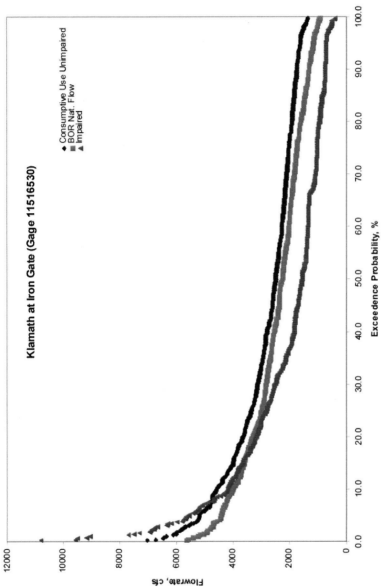

FIGURE 5-5 Annual flow duration plots at Iron Gate for the observed and estimated flows based on the USBR NFS and level-pool, consumptive-use-based methods.

SOURCE: Hardy et al. 2006a, Figure 7. Reprinted with permission; copyright 2006, Utah State University.

Hydrodynamic Modeling

Hardy et al. (2006a) used a high-resolution integrated digital terrain model (derived from aerial photographs and hydro-acoustic data), and results from substrate and vegetation mapping as input to a two-dimensional, quasi-three-dimensional hydrodynamic modeling program developed by the USGS (Nelson 1996, Thompson et al. 1998, Nelson et al. 1995, McLean et al. 1999, Topping et al. 2000) (Figure 5-6). The name of the flow model is the Multi-Dimensional Surface Water Modeling System (MD-SWMS). The MD-SWMS computes down-stream and cross-stream components of velocity and boundary-shear stress based on finite-difference solutions to the depth-averaged Navier-Stokes equations for turbulent flow. Details of the model and its application are given in Lisle et al. (2000), and at the MD-SWMS site: http://wwwbrr.cr.usgs.gov/projects/SW_Math_mod/Op-Models/MD_SWMS/index.htm. The MD-SWMS model is used by Hardy et al. (2006a) to simulate the distribution of depth and velocity in each of the study sites for a range of flows. Computational meshes used in hydraulic simulations contained nodes every 1.6 m (5.25 ft) across the river and 1.7 m (5.58 ft) in the longitudinal direction. Hardy et al. (2006a) modified the

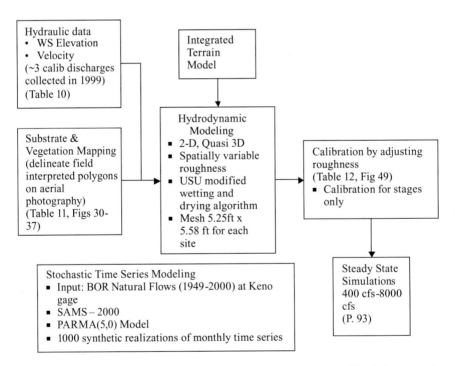

FIGURE 5-6 Overall data flow diagram of the hydrodynamic and hydrologic modeling step in the IFS Phase II study.

wetting and drying algorithm in the original USGS program to improve its performance.

The scale of the study sites and size of the model grid is such that each model run generates several hundred thousand values of depth and velocity, spatially distributed across the channel. These distributions indicate not only the percentage of suitable habitat available to individual species and life stage at that flow, but the modeled values of depth and velocity are assigned to specific locations and translated into area. The strength of this approach is that it allows for explicit testing of the habitat-suitability model output, assuming that independent presence and absence observations of fish distribution through a modeled reach can be made and compared with suitable versus unsuitable wetted areas.

Hardy et al. (2006a) developed a hydrodynamic model for each study site. For calibrating the models, they collected three sets of water-surface elevations at three different discharge levels for each study site. Model calibration consisted of matching observed and predicted water-surface elevations. Predicted water-surface elevations were manipulated by adjusting spatially variable model roughness coefficients. Differences between observed and predicted water-surface elevations were generally between 1 cm and 5 cm. Calibration of the model velocities is not required. An assessment of differences between observed and predicted velocities was conducted by Hardy et al. (2006a, Appendix F) and it was concluded that the differences were within expected and acceptable ranges. The calibrated model simulated flows ranging from 400 cubic feet per second (cfs) to 8,000 cfs at Iron Gate Dam. Hardy et al. (2006a) used output from the simulations in habitat modeling.

Stochastic Modeling

In most cases, river flows have significant periodic or seasonal behavior in the mean, standard deviation, and skewness (Tesfaye et al. 2006). In addition to these periodicities, they show a time correlation structure (autocorrelation), which may be either constant or periodic. For river flows, autocorrelation usually arises from the effect of surface, soil, and ground storage, which causes the water to remain in the system through subsequent time periods. Periodic Autoregressive Moving Average (PARMA) models are an important class of models that explicitly account for seasonal fluctuations in the mean flow, flow standard deviation, and flow autocorrelation. The notation used to designate the order of a PARMA model is PARMA (p,q), where p and q are integers that represent the model order. For example, a PARMA model that captures watershed processes having a lag time of up to 5 months would be designated as a PARMA $(5,0)$ model.

Hardy et al. (2006a) used a PARMA $(5,0)$ model (Salas 1993) to derive 1,000 synthetic monthly flow series at Keno. Hardy et al. (2006a) then added unimpaired accretions to the Keno flow series to obtain flow series for Iron Gate Dam. For deriving the synthetic Keno flow series, Hardy et al. (2006a) used flows reported by USBR (2005). The primary reason for generating the flow series was to estimate "uncertainty bands" for the USBR natural flow estimates at various exceedance frequency levels (Hardy et al. 2006a, p. 85). In addition, Hardy et al. (2006a) used 100 of the synthetic flow series to compute physical habitat time series at the R-Ranch study site (that is, Iron Gate Dam).

Hydraulic-Habitat Model Development and Testing

Hardy et al. (2006a) included a selected number of salmonid species and life stages in the main stem for habitat assessments: steelhead (fry and 1+ [juveniles]), Chinook salmon (spawning, fry, and juvenile), and coho salmon (fry and juvenile). Using field-data collection and data-reduction methods described by Hardin et al. (2005), Hardy et al. (2006a) developed site-specific habitat-suitability criteria for Chinook salmon spawning, fry, and juveniles, coho salmon fry, and steelhead 1+ (juveniles). Because of lack of adequate field data, literature-based habitat-suitability criteria were used for coho juvenile and steelhead fry. Using the individual habitat-suitability indices for depth, velocity, substrate, and composite escape cover, Hardy et al. (2006a) developed composite suitability indices for the selected species and life stages (see Figure 5-7). Composite escape cover is a weighted index of nodes around a particular location. It is a function of the type of escape cover and binary variables representing distance to escape cover, escape depth threshold, and an escape-cover velocity threshold.

The hydrodynamic variables simulated for mesh nodes at the intensive study sites at representative discharges were used to compute composite suitability indices for each site, species, and life stage. Hydrodynamic modeling results for the discharge closest to that observed during fish field data collection periods were interpolated at 2-foot mesh intervals and used to compute composite suitability indices. Field observations of fish locations were plotted on GIS maps of computed composite suitability indices and orthophotos to compare habitat modeling predictions with field observations.

Habitat Modeling

In the habitat-modeling phase of the study (Figure 5-8), Hardy et al. (2006a) converted the composite suitability indices described above into a combined measure commonly known as the weighted usable area (WUA) to characterize the quality and quantity of habitat in terms of usable area per 1,000 linear feet of stream. Hardy et al. (2006a) used steady-state

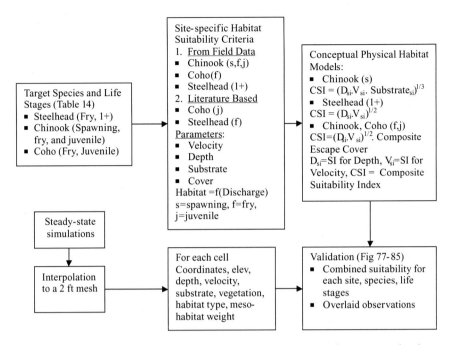

FIGURE 5-7 Overall data flow diagram of the habitat-suitability criteria development and validation step in the IFS Phase II study.

hydrodynamic simulations to produce percent maximum habitat WUA as a function of discharge for each study site, species, and life stage. Scaling study-site habitat values to the larger physiographic reach level is a challenging task. To do this, Hardy et al. (2006a) overlaid the mesohabitat characterization provided by USGS/USFWS for the study site and assigned each computational node of the study site to a mesohabitat type (for example, their Figure 121). To scale up from the site to the reach level, Hardy et al. (2006a) assigned a weight to each computational node based on the surface area in the reach of the mesohabitat type at that node. In other words, the relative area in each mesohabitat type at the reach level was used to weight the habitat data at each computational node in the study site. Using that weighting, reach-level WUAs could be computed directly from study-site WUAs. Based on information provided to the committee after release of the report, the weighting was computed as

$$\text{Weight} = \frac{\left(\dfrac{\text{HabTypeArea\_Reach}}{\text{HabTypeArea\_Site}}\right) * 10^5}{\text{HabTypeArea\_Site}},$$

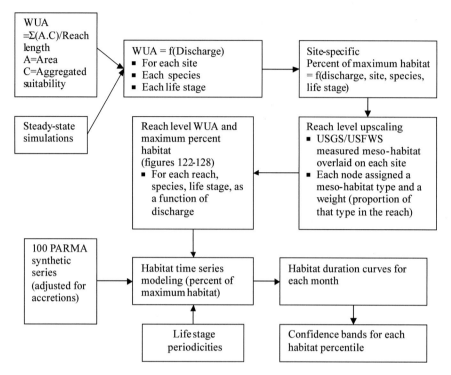

FIGURE 5-8 Overall data flow diagram of the habitat modeling step in the IFS Phase II study.

where HabTypeArea is defined as the area of the habitat for a particular mesohabitat type.

Instream Flow Recommendation and Justification

As the final step, Hardy et al. (2006a) combined the monthly flow-frequency curves generated for three hydrologic scenarios (existing, UBSR natural flows, and consumptive-use based), an analysis of historical peak flows and the percent maximum habitat frequency curves and their confidence bands (previous step) to develop instream flow recommendations (Figure 5-9).

Using NRC (2005a) as the primary source, Hardy et al. (2006a) discussed the following components of an instream flow recommendation: (1) overbank flows, (2) high-flow pulses, (3) base flows, (4) subsistence flows, and (5) ecological base flows, although this final component was not part of NRC (2005a). Hardy et al. (2006a) consider overbank flows necessary

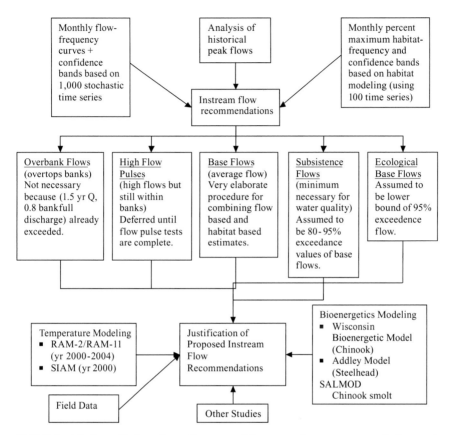

FIGURE 5-9 Overall data flow diagram of instream flow recommendation and justification step in the IFS Phase II study.

for maintenance of alluvial channels to be discharges exceeding 80% of bankfull discharge at a particular location based on a study in gravel-bed rivers (Schmidt and Potyondy 2004). Assuming that dams and reservoirs in the main-stem Klamath do not have adequate storage capacity to affect peak flows, and supported by an analysis of historical data, Hardy et al. (2006a) argue that no prescription of overbank flows is required; however, the frequency and duration of such peak flows were not analyzed. In addition, Hardy et al. did not prescribe high-flow pulses. While recognizing the importance of these flows, they reasoned that state and federal agencies are evaluating the use of a water bank for generating pulse flows and deferred the specification of these flows until that work is completed.

For base flows, Hardy et al. (2006a) used an elaborate procedure to combine flows based on the natural flow paradigm and flow-habitat functions based on reach-level physical-habitat modeling. The exact details of the procedure were not completely clear, but based on the descriptions in Hardy et al. (2006a), the following procedure was used.

Hardy et al. (2006a) appear to have recognized two key points:

• The flow estimates for a particular exceedance frequency and those that correspond to maximum percent habitat will not be the same. A procedure is needed to integrate these two estimates.

• A flow estimate that maximizes the habitat for one species or life stage might not maximize the habitat for other species or life stages. In other words, there is no single flow recommendation that can maximize the habitat for all species and life stages at the same time.

Two flow estimates were derived and combined for developing instream flow recommendations:

1. Flow-based estimates were directly based on USBR natural flow estimates, on consumptive-use flow estimates produced in an earlier study (referred to on p. 32 of Hardy et al. 2006a, but not identified), and on existing flows reported by the USGS.

2. Habitat-based estimates were based on the USBR natural flow estimates. Hardy et al. (2006a) used the habitat-suitability-criteria modeling, and produced a set of habitat-based flows through a series of manipulations of the modeling results.

Hardy et al. (2006a) produced the final flow recommendations as the average of the flows produced in the two estimates described above. Those estimation procedures are described in more detail below.

For flow-based estimates, Hardy et al. (2006a) plotted flow versus percent exceedance for each month and developed a linear relationship for flow and percent exceedance. This relationship is of the form:

$$Q = ax + b,$$

where x denotes the estimated percent exceedance, and Q is the discharge. The coefficients a and b vary monthly. The above relationships were developed for the USBR natural flows, consumptive-use-based flows, and the current managed flows reported by the USGS. Based on these relationships, a flow corresponding to each percentile was calculated (Q_{Nat}, Q_{CU}, and Q_{Exist}) corresponding to various exceedances, for example, for 90% exceedance. Hardy et al. (2006a) then computed estimates of the upper and lower limits of this percentile using the stochastic time series that were generated from the PARMA model fitted to the USBR naturalized flows.

The bounds were computed as follows (as an example, for 90% exceedance of the USBR natural flow):

$$Q_{Nat,L}^{90} = f_{lower} * Q_{Nat}^{90}$$

and

$$Q_{Nat,U}^{90} = f_{upper} * Q_{Nat}^{90},$$

where the subscripts L and U denote lower and upper bounds, respectively. The multiplication factors in the above equations were computed using the confidence bands computed from 1,000 series of stochastically simulated flows based on the USBR natural flows. The flow-based estimate of the instream flow recommendation, for example, for the 90% exceedance level, was then determined by the following formula:

$$Q_{Flow}^{90} = \min(\max(Q_{Nat,L}^{90}, Q_{Exist,}^{90}), Q_{Nat,U}^{90}).$$

It appears that the consumptive-use-based flows did not play a role in the final instream flow recommendations.

For physical-habitat-based flows (from habitat modeling), a procedure similar to the flow-based estimate was used. A habitat versus percent exceedance curve was generated from the habitat modeling using 100 stochastic time series. The mean habitat value corresponding to a particular percent exceedance was then used to compute a corresponding pair of flow estimates, Q_L and Q_U. For a given habitat value, the habitat-suitability curve can yield two optimal flow estimates.

Hardy et al. (2006a) then used an elaborate filtering procedure to determine the habitat-based instream flow recommendation, Q. The above calculations were carried out for each species and life stage. Then the final habitat-based instream flow recommendation was computed as the geometric mean:

$$Q^{Hab} = (\prod Q_i)^{1/n},$$

where subscript i refers to the species or life stage.

The final instream flow recommendation for a particular exceedance frequency was computed as the average of flow-based and habitat-based flow values:

$$Q_{Final} = 0.5 \left(Q_{Rec}^{Flow} + Q_{Rec}^{Hab} \right).$$

The above elaborate procedure was designed to integrate the flow-based and the habitat-based instream flow recommendations into a single value that provides an "average" flow condition. This average is intended to satisfy the needs of all species and life stages as much as possible, but it cannot be optimal for each (or any) of them.

The only way to justify the instream flow recommendations developed using the above procedure is to conduct habitat modeling to determine if such recommendations indeed result in better habitat than the current flow regime provides. Hardy et al. (2006a) conducted limited habitat modeling to demonstrate this, but more probably is needed.

Hardy et al. (2006a, p. 181) state that the final flows recommendations for a location below the Iron Gate Dam (Table 27, Hardy et al. 2006a) provide improved habitat: "70% of the habitat values were within the lower and upper limits of the expected habitat variability for a particular stage and month exceedance level. Of the 30% of cases where the recommendation fell below the lower limit for habitat, 88.6% still showed an improvement of approximately 20% on average compared with the existing conditions."

Hardy et al. (2006a) discussed subsistence flows, defined as flows necessary to maintain adequate water-quality conditions, and assumed to be flows corresponding to monthly 80-95% exceedance frequencies. Ecological base flows, defined as those at which further reductions would result in unacceptable levels of risk to aquatic resources, were calculated as the lower bound of the flow corresponding to monthly 95% exceedance levels.

Temperature Modeling and Robustness of Flow Recommendations

The flow recommendations were based primarily on hydraulic modeling of physical habitat, the NFS, and base flows as described above. Equally important to the hydraulic habitat aspect is the temperature regime. To examine the robustness of the hydraulic habitat-based flow recommendations, Hardy et al. (2006a) conducted independent analyses of water temperature, bioenergetics of Chinook salmon and steelhead, and population dynamics of Chinook salmon. Hardy et al. used the RAM-2/RMA-11 temperature modeling conducted by Watercourse Engineering (Deas et al. 2006) for 2000-2004 and temperature modeling using monthly flows with the Systems Impact Assessment Model (SIAM) (Bartholow et al. 2003) to evaluate temperature effects on anadromous species growth and production.

Hardy et al. (2006a) conducted bioenergetics modeling for Chinook salmon and steelhead, concluding "that the higher growth rates and increased availability of physical habitat associated with our flow recommendations result in a net benefit to Chinook and steelhead rearing conditions. . . ." An independent analysis simulating growth, movement, and survival of outmigrant Chinook salmon was conducted using the fish population model SALMOD, as it had been incorporated within the SIAM model by USGS (Bartholow et al. 2003). Hardy et al. (2006a) conclude that these independent analyses "demonstrate the proposed flow regimes benefit all life stages of anadromous (Chinook salmon) species during all periods of the year including upstream migration, spawning, rearing, up-stream passage, and out-migration compared to existing conditions." The final flow recommendations were further justified by citing field data and studies conducted elsewhere.

EVALUATION OF PHASE II INSTREAM FLOW STUDY

Site Selection

Hardy et al. (2006a) classified the main-stem Klamath River below Iron Gate dam into five "homogeneous" study reaches based on the junctions of four major tributaries to the main stem. Among other things, the purpose of the stratification was to delineate sections of the river that function in a similar manner in terms of flow volumes, overall channel characteristics, species and life-stage distributions. Although the practical necessity of stratifying the river into reaches is understandable, an analysis justifying the assumption of homogeneity between reaches should have been included in the report.

Hardy et al. (2006a) selected eight study sites (investigators later dropped one site due to unsatisfactory data) within the five study reaches. The eight sites were selected based on input from a technical review team, and representatives of various federal, state, and tribal agencies. The sites were considered to be broadly characteristic of individual reaches and were established in some locations to coincide with other studies of water quality and water temperature.

There is no reason to think that the sites are unrepresentative of the respective study reaches; however, given that the flow recommendations hinge on modeling results from these eight sites, it is important to establish whether they are representative of the reaches. Most of the questions concerning site representativeness could be addressed with field measurements of basic channel properties (width, depth, slope, and substrate composition) at intervening locations (perhaps every few kilometers) between the study sites.

Independently, the USFWS, USGS, and Yurok Tribe conducted field-based mapping of mesohabitat types (Hardy et al. 2006a, Table 2) from Iron Gate Dam to the estuary. The study sites and reaches contain different mixtures of mesohabitat types, which make scaling the study-site results to the reach level difficult. Although it is not possible to reanalyze the data extensively at this stage, an analysis to demonstrate a logical stratification of segments, mesohabitats, and study sites (microhabitats) should be provided (Bovee 1982; also see Figure 1-2 in Bovee et al. 1998).

Hydrology and Hydraulics

Using Monthly Flows to Determine Recommended Flows

Most of the hydrologic and hydraulic analyses and models used by Hardy et al. (2006a, p. 24) involved standard and widely used methods. Nonetheless, several of the analyses could be improved, better justified, or lack substantiation. In the following paragraphs, the committee comments on the hydrologic and hydraulic analyses conducted for the IFS.

Standard practice for determining recommended flows includes use of daily (or at least weekly) hydrologic records (Annear et al. 2002, p. 131). Flow recommendations for the Trinity River, California, for example, were developed using daily and weekly hydrologic information (USFWS/HVT 1999). In discussing hydrologic alterations in the Klamath River basin, Hardy et al. (2006a, p. 24) indicate that they considered annual, monthly, and daily flows. Unfortunately, the important data sets provided to Hardy et al. were based on different time steps. The hydrology data from the NFS were monthly, and temperature data from the RMA (Resource Management Association, discussed later in this chapter) were daily. Integration of hydrology, temperature, and habitat as inputs to generate fish-population time series requires a common time step.

To underscore the detrimental effect of making flow recommendations for the Klamath River based on monthly versus daily mean flow values, the committee considers flow at two locations: the Klamath River below Iron Gate Dam and the Williamson River near Chiloquin, an upper-basin tributary to the Klamath River. There are no major impoundments above the mouth of the Williamson River, and it is the largest water source for the Upper Klamath Lake (NRC 2004a, p. 26). Flows at the mouth of the Williamson River are affected by privately managed irrigation diversions, but given the large total flow in the Williamson, the hydrograph has predominantly natural features. In contrast, a great deal of water management occurs at Iron Gate Dam; that management and the retention of water in reservoirs on the Lost River and in Upper Klamath Lake have the potential to alter the hydrograph extensively.

Figure 5-10 shows daily and monthly mean flows for the Klamath River below Iron Gate Dam and Williamson River near Chiloquin in 1993, a year of near-average water availability (NRC 2004a). In both cases, the variability of the daily discharges between March and June indicate that monthly flows are insufficient for assessing flow-duration in the Klamath River basin

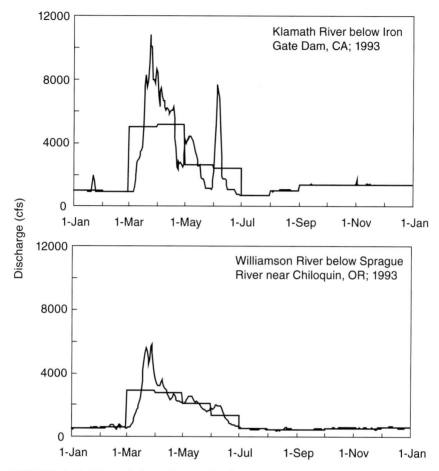

FIGURE 5-10 Mean daily and monthly flow of the Klamath River below Iron Gate Dam and Williamson River near Chiloquin in a year of near-average water availability.
SOURCE: Data from U.S. Geological Survey.

as was done for developing Figure 5-5. Also because of variability in the daily flows, the use of monthly flows for assessing habitat, water quality, and temperature is of dubious value. Figure 5-10 illustrates that it is not possible to identify important hydrologic regime components (Table 5-1) from monthly average flows because monthly averaging obscures the magnitude and duration of high-flow pulses and overbank flows.

The committee concludes that Hardy et al. (2006a) should have used daily flows or at least weekly flows for making instream flow recommendations, because monthly time steps are likely to produce erroneous results.

To address this shortcoming, the committee recommends that consideration be given to stream flow disaggregation modeling as a means for obtaining daily stream flow data while preserving the statistical attributes of the estimated monthly flows. Bartolini and Salas (1993), Kumar et al. (2000), Tarboton et al. (1998) and others offer methods for disaggregating hydrologic time series. Alternatively, the committee recommends considering the development of deterministic, daily watershed models for each of the major sub-basins of the Klamath River basin. An example of this type of modeling is the Natural System Model (Fennema et al. 1994) for simulating the hydrologic response of the natural Everglades in south Florida.

Routing Natural Flow Estimates

Two studies were conducted by the USBR to estimate natural flows in the Klamath River basin: the first used a consumptive-use, level-pool routing method and the second used a water-budget approach. In the consumptive-use, level-pool routing study, the USBR used accretions currently used in their operations models for the Klamath Project to route flows from Upper Klamath Lake to Iron Gate Dam. Since the USBR accretions represent impaired accretions, Hardy et al. (2006a, p. 33) developed and used their own accretion estimates to route the USBR flow estimates for Keno gauge, from Keno gauge to Iron Gate Dam. It is appropriate that Hardy et al. used their own accretion estimates, because using impaired accretions to estimate natural flows is not logically consistent.

Stochastic Modeling and Normalized Flows

Hardy et al. (2006a) used a PARMA 5,0 stochastic model to generate synthetic monthly flows at the Iron Gate Dam. They generated synthetic flows to account for uncertainty in the natural flow estimates and to compute physical-habitat time series. The stochastic modeling approach that Hardy et al. used has the following limitations:

- The natural flow estimates for Iron Gate Dam have many uncertainties due to approximations used in the NFS (USBR 2005) and for estimating unimpaired accretions (Hardy et al. 2006a). Furthermore, the USBR estimated natural flows using conservative assumptions so that the true natural flow would not be underestimated (T. Perry, USBR, personal communication, 2006). As a result, there is no guarantee that the PARMA modeling can reasonably address uncertainties in the monthly flow series. A formal error analysis of the NFS would be required to develop uncertainty bands for the estimated natural flows.
- If the consumptive-use-based natural-flow estimates for Iron Gate

Dam are reasonably accurate, then the stochastic monthly flow series may represent future realizations of natural flows. However, a PARMA (5,0) model captures only watershed processes that have a lag time of up to 5 months, and therefore, it does not integrate annual autocorrelation, which likely is present because annual autocorrelation due to long-term memory is a common characteristic of watersheds with significant groundwater storage such as the Klamath basin.

•   It appears that the PARMA model was fit without any normalizing transformation of the raw "naturalized" data. This may partially explain why the model is not able to reproduce small flows, that is, flow with exceedance levels greater than about 75%. Although Hardy et al. claim to have circumvented this problem by confining their use of the stochastic flow series to considering variability of the mean flow, their approach does not guarantee that the results are reasonable.

If stochastic hydrologic modeling is to be used, it is necessary to consider the stochastic nature of hydrologic time series at multiple sites, not just at one location such as Iron Gate Dam. Tributary flows, although probably highly correlated with main-stem flows, are not completely correlated with them or accurately predictable based on main-stem flows alone. Effective decision making for realizing future "naturalized" flows accounting for stochastic dependence across key points in the basin requires the creation of a multi-site, periodic, stochastic hydrologic model.

Physical Processes, Geomorphology

Hardy et al. (2006a) note that "an important aspect of providing flow recommendations that will protect the important physical, chemical, and biological processes of the river corridor relates to the dynamic processes associated with channel and riparian maintenance flows," but they dismiss this critical component of a comprehensive instream flow prescription by referencing analyses performed by PacifiCorp (2004), Ayres Associates (1999), and others. According to Hardy et al. (2006a), PacifiCorp's analyses of the Klamath River indicate "the basic planform of the river at the reach scale has not changed over the last 50+ years" and "most alluvial features and associated riparian vegetation communities remain dynamic." Hardy et al. implicitly assume, based on the analyses conducted by PacifiCorp and others, that the present channel bathymetry will not change under the recommended instream flow regime. Although the present channel form may not have changed over the past 50+ years, that is no guarantee that it will not change if the Klamath River's flow regime is significantly altered. Regardless, to be effective, instream flow recommendations should explicitly discuss flows required to maintain channel dynamics.

Physical Habitat Simulation

A principal objective of Hardy et al. (2006a) was to develop simulation models for evaluating spatiotemporal relations between discharge and physical-habitat suitability in different reaches of the Klamath River. The habitat simulations in Hardy et al. follow the basic structure of the PHABSIM output and habitat time series analyses to identify fish species life-stage habitat constraints ("habitat bottlenecks") as part of IFIM (Bovee et al. 1998; see Chapter 3 for discussions of PHABSIM and IFIM), including a set of software programs developed by the USFWS/USGS (Bartholow and Waddle 1986, Waddle 2001). This approach attempts to simulate how changes in stream flows affect the potential micro-habitat usability in time and space for different species of fish at different life stages. This type of modeling essentially simulates the preference-avoidance behavior of specified fish life stages to changes in the hydraulic aspects of stream flow. Micro-habitats are specified in terms of differences in physical conditions within the river channel (the distribution of flow, depth, velocity, substrate, and cover characteristics), which fish use according to their specific needs. Since these conditions are rarely known at more than a few locations on a river (and perhaps at only few flow levels), the strategy in developing flow recommendations for the benefit of fish or other aquatic organisms is to sample sufficiently to simulate changes in habitat suitability at a several key locations over a range of flows, and then to expand these relations to larger physiographic river segments.

Biology: Fish Life Histories and Instream Flows

Species and Life Stages Considered

The main-stem Klamath provides habitat for 19 species of native fishes (13 of which are anadromous and 2 amphidromous) and 17 nonnative fish species (NRC 2004a, Chapter 2). The river supports eight tribal trust species that are anadromous: steelhead, coho salmon, Chinook salmon, green sturgeon, white sturgeon, coastal cutthroat trout, eulachon, and Pacific lamprey (Hardy et al. 2006a). The flow recommendations provided by Hardy et al. are supposedly "made in consideration of all the anadromous species and life stages on a seasonal basis and do not focus on specific target species or life stages (i.e. coho)" (Hardy et al. 2006a, p. ii). In fact, the flow recommendations reflect a focus on usable hydraulic habitat for several life stages of three species: steelhead, Chinook salmon, and coho salmon (Hardy et al. 2006a, Table 14). Hardy et al. note that this is due to the limited data available for the other species. This limited focus calls into question whether or not the recommended flow regime will "provide for the long-term protection, enhancement, and recovery" (Hardy et al. 2006a, p. ii) of all aquatic resources within the main-stem Klamath River.

One argument for considering a "natural flow regime" is that it better reflects the requirements of the assemblage of species rather than individual species of concern (Poff et al. 1997). Hardy et al. (2006a) used the NFS as a starting point for flow recommendations (for example, Table 25, Hardy et al. 2006a). Use of this approach was recommended by the NRC (2004a, p. 300). Yet Hardy et al. offer no consideration of the consequences of recommended flows on any of the other 16 species of native fishes, despite their cultural significance for Native Americans (for example, eulachon), or on any other vertebrate or invertebrate species dependent on the Klamath River main stem.

Hardy et al. agree with others (for example, NRC 2004a) that tributary conditions are important for coho salmon fry and juveniles, but justify the inclusion of these life stages when developing main-stem flow recommendations because spawning coho salmon were observed in the main channel, and 4,000 coho salmon fry were captured in traps in 2002. Furthermore, coho salmon smolts use the main stem for several days during out-migration (Stutzer et al. 2006). Using capture efficiencies from Chinook salmon smolts, Hardy et al. estimate that 1.2 million coho salmon fry passed the main-stem traps. Although this is an interesting extrapolation, given the life history of coho salmon, the significance of these fish to the overall population of the species is unknown. Also unknown is whether they are out-migrating to the estuary, will remain in the main-stem Klamath, or will find tributary habitat to complete their juvenile life history before smolting.

Hardy et al. (2006a) describe the life stages and associated spatial and temporal use of main-stem habitat for coho salmon, Chinook salmon, and steelhead as "priority species for the flow recommendations." In their Tables 15 and 16, Hardy et al. further distill the complex life-history needs with a synthesis of the species' needs by month for a bridge between biological needs and a monthly flow assignment. The tables represent a coherent and appropriately simplified picture of a complex amalgam of species needs.

Hardy et al. (2006a) used a standard approach to develop the percentage of maximum WUA available to each species and life stage under different flow conditions, and they enhanced the analysis by developing a modeling approach to incorporate access to escape cover. The analysis incorporated both distance to and minimum depth requirements for access to escape cover for juveniles and fry. Hardy et al. argue that use of these behavior-based models improved the model's ability to predict habitat requirements of Chinook fry. They report 80% of fish were observed where the combined suitability was 0.75-1.00, and only 5% were observed where suitability was less than 0.25. This modeling approach represents an advancement of the field. The most extensive data available to Hardy et al.

(2006a) were for Chinook fry and juveniles, and those data were also used for coho juveniles and steelhead fry. Limited field data were available on escape cover for coho fry and steelhead juveniles.

The NRC (2004a) criticized an earlier draft report (Hardy and Addley 2001) for basing coho habitat requirements on those of Chinook. In Hardy et al. (2006a), habitat-suitability requirements are based on a more extensive field study (Hardin et al. 2005) for coho fry; spawning, fry, and juvenile Chinook; and steelhead juveniles (1+). Values derived from the literature were used for coho juveniles and steelhead fry. This treatment represents an improvement over the earlier study.

Testing Habitat Model Output

Hardy et al. (2006a) acknowledged the complexity of fish habitat-suitability criteria in that "habitat use may change with fish size, season, temperature, activity, habitat availability, presence and abundance of competitors and predators, discharge, and changes between years. . . .These factors underscore the importance of validating the HSC [habitat-suitability criteria], especially in terms of the habitat modeling results." The habitat-suitability model was tested by visually comparing predicted fish distributions with observations at several study sites under different flow conditions. In some cases observations were extremely limited (for example, two coho juveniles). Based on these visual comparisons for each species and life stage, Hardy et al. conclude that the modeled habitat availability is adequate for instream flow assessments. For example, for steelhead juveniles, they conclude, "there is generally good agreement between predicted and observed habitat utilization over different flow rates and at different stations" (Hardy et al. 2006a, p. 147). These conclusions would be stronger had they been based on a statistical analysis (see Chapter 2) across sites and considering the frequency of observations in habitats with different values for habitat-suitability criteria. Even a simple chi-square analysis to determine if the likelihood of occupancy was greater than expected by a random distribution would have provided greater confidence in the habitat-suitability modeling. This type of analysis has been done in studies considering the transferability of habitat-suitability criteria from one river to another (for example, Thomas and Bovee 1993, Freeman et al. 1997).

The committee has similar concerns about the approach used to evaluate the impact of recommended monthly flows on the percent of maximum WUA for each species and life stage. The conclusion—"at most study sites the habitat availability for life stages of anadromous species are maximized at seasonal flow ranges estimated in the 'natural flow' study below Iron Gate Dam" (p. 155)—is not quantified or supported by figures or tables clarifying what is meant by "most study sites" or "maximized." Habitat time series and monthly habitat-duration curves were provided that compared natural flows, existing flows, and the imposed flow recommendations. These analyses should be expanded and incorporated in the Hardy et al. (2006a) report to further support and justify conclusions as to the likelihood that the recommended flows will protect, enhance, or recover the anadromous salmonid fishes.

A more explicit analysis of the life history and habitat time series to identify potential habitat bottleneck(s) and the life stage(s) most influenced would be most helpful. These analyses are important components of a comprehensive IFIM study. This was addressed to a limited extent in the Phase II report through the SALMOD model to illustrate the efficacy of recommended flows as compared with existing and natural flows. The Hardy Phase II study went further in conducting these analyses than most other studies of riverine habitats and instream flow assessments. However, modeling the use of fish habitat and populations is scientifically challenging, as is the management of instream flows. Additional analyses using daily time steps and rigorous model testing would enhance confidence in these results.

Weighted Usable Area in Recommending Instream Flows

Although standard practices were generally used by Hardy et al. (2006a) in their assessment of habitat suitability, there can be fundamental problems with the standard WUA approach when used in isolation from other important components of the flow regime (for example, temperature, water quality, sediment transport, and channel movement). The objective of Phase II was to recommend flows that provide protection, enhancement, and recovery of anadromous fish populations; yet WUA is at best an indirect indicator of population status. WUA is a measure of habitat suitability within the confines of the simulation of usable versus unusable habitat area over time and space; it predicts how likely a habitat patch is to be occupied or avoided by a species life stage at a given time, place, and discharge. Habitat-suitability modeling has inherent limitations that have received widespread recent attention in general (for example, Garshelis 2000, Burgman et al. 2001) and in reference to instream flow assessments (for example, Orth 1987, EPRI 2000, Railsback et al. 2003). Among the most relevant of these limitations are the following:

1. The fundamental assumption that populations respond in proportion to the availability of highly suitable habitat versus unsuitable habitat is not well documented. Unlike the Phase II study, most studies omit the important step of model calibration and of testing this fundamental assumption. Fish populations are limited in part by factors (for example, food

availability, predators) other than physical habitat. Any physical-habitat study should investigate the extent that habitat may be limiting under the specific flow regimes that are considered for the stream in question. The IFIM comprehensive study process emphasizes this question and provides guidelines for assessing the role of physical habitat in comparison with other limiting factors.

2.   Competition within and among species for habitat can result in misleading habitat-suitability models, particularly for the Klamath River. For example, Railsback et al. (2003) used individual-based models and found juvenile salmonid population responses to flow changes to be the opposite of those predicted with untested habitat-suitability modeling. Habitat created for small fish can be occupied instead by larger fish or fish of a different species. The use of independently derived fish distributions, as compared with model output, provides some support to the validity of the fry and juvenile salmon habitat model for use on the Klamath River, but these comparisons require more rigorous statistical analysis.

3.   Habitat suitability can be strongly affected by many factors that vary over time. The most beneficial velocities and depths are known to vary with factors such as fish size, life-history status, water temperature, turbidity, food availability, competition for food, and predation risk. Developing habitat-selection models from field data that address all these factors would be extremely difficult. Omitting them ignores important ecological interactions. Testing habitat-suitability model output with independently derived observations of fish distribution is necessary for acceptance or rejection before incorporating into analyses leading to instream flow prescriptions.

4.   Habitat-suitability models produce different results for different life stages and different species. Use of physical habitat versus flow relations in isolation provides no meaningful prediction of overall effects on a species or community. For example, at a specific stream location a change in flow might indicate a doubling of suitable habitat area for salmon fry but one-half the suitable habitat area for larger juveniles. These results by themselves provide no meaningful way to predict the overall effect on production or the population status of stream discharge. Time series of stream hydrology translated to distribution of usable habitat over time and space coupled with species periodicity is necessary to identify potential physical habitat limitations (bottlenecks). Such analyses may identify the life stage(s) most vulnerable and the timing (seasonal or inter-annual) of likely habitat-limiting events.

There is extensive ecological literature on habitat-selection modeling, which indicates that simple selection of flow recommendation from a static set of WUA versus flow curves is not considered a credible approach (for example, Marthur et al. 1985; Shirvell 1986, 1994; Osborne et al. 1988;

Gan and McMahon 1990; Elliot 1994; Castleberry et al. 1996; Ghanem et al. 1996; Williams 1996; Lamouroux et al. 1998). It has been long recognized that WUA versus flow relations alone lack biological meaning. Instead, the flow-habitat relations as a component of IFIM are intended as input for evaluation of habitat usability through time and space using time-series and effective-habitat analyses (Stalnaker 1979; Stalnaker et al. 1995; Bovee et al. 1998; NRC 2005a, 2005c). There is no justifiable way to base flow-management recommendations on habitat-flow relations alone. This limitation of WUA has been recognized for more than 30 years, but WUA continues to be inappropriately used by some as a shortcut approach for selecting flat-line "minimum flows." Appropriate instream flow recommendations include both intra- and inter-annual variability (Stalnaker et al. 1995, Annear et al. 2004, NRC 2005a). The issue is properly addressed by Hardy et al. (2006a) in the Flow Recommendation Methodology section, where a sequence of steps was taken to translate WUA results into habitat time series and the development of flow recommendations.

Alternatives to a simple flow-WUA approach are provided in the ecological modeling literature. For example, Manly et al. (2002) recommended modeling the density of animals, which can be done using methods similar to those used for WUA. Density has a clear biological meaning that can be translated into specific outputs, such as total numbers of fish.

The use of habitat–time series, coupled with the life-history periodicity and some estimate of habitat carrying capacity, is the recommended approach within IFIM. This is incorporated into quasi-population analyses using effective-habitat analyses (Bovee 1982), or—for salmonid fishes—the population model SALMOD (Bartholow et al. 2002). Both methods use temperature and flow time-series analyses along with physical habitat to identify potential bottlenecks imposed by alternative flow regimes (Bartholow and Waddle 1986, Bartholow et al. 1993, Bovee et al. 1998).

Assessing the Relevant Environmental Variables

Critical to producing a good measure of habitat selection is the choice of driving environmental variables. Leaving out important variables reduces a model's predictive ability, yet including too many variables can cause problems such as over-fitting (attempting to be too precise in fitting a model to the data) and unnecessarily high uncertainty. A modern approach is to select a variety of habitat variables likely to affect habitat selection, evaluate those in the field by observing animals, and then use statistical analysis to determine which variables are important enough to include in a model (for example, Manly et al. 2002). Typically, information-theoretical approaches, such as Akaike's information criterion, are used to determine which habitat variables should be in a model. The scientific literature confirms that depth, velocity, substrate, cover temperature, and flow dynamics are major determinants of the distribution and abundance of stream fishes, especially salmonids (Annear et al. 2002). Hardy et al. (2006a) assumed, a priori, that physical habitat (depth velocity, substrate type, and distance to cover), temperature, and flow dynamics are the most important variables. Nonetheless, for a specific locale, evidence needs to be provided to justify those assumptions and to justify the conclusion that the recommended flow prescriptions would best provide for these variables in this locale.

Modeling Fish Growth

Bioenergetics models were used to explore the impact of recommended flows on fish growth as a function of temperature. The temperature used in the models was daily "fish mean temperature," which is the daily mean temperature plus 40% of the difference between daily mean and maximum temperatures. Hardy et al. (2006a) cite several studies supporting this approach. The NRC (2004a) argued for the critical importance of diel minimum temperature, but Hardy et al. argue that the main importance of diel minima is to reduce the daily mean temperature. The recommended flows result in lower maximum and mean temperatures, but higher daily minimum temperatures; so if diel minimum temperature, rather than "fish mean temperature," is the critical determinant of fish health, these recommendations are flawed. This is further evidence that daily hydrologic flow data are needed.

The Wisconsin Bioenergetics Model (Hanson et al. 1997) was used for Chinook salmon, and a model developed for rainbow trout above Iron Gate Dam (Addley 2006) was used for steelhead (the anadromous form of rainbow trout). Growth is a function of temperature but also of food availability. The latter was determined to be 41.5% of maximum consumption by fitting the model to observations of Chinook growth (Hardy et al. 2006a, Figure 151). The use of thermal refugia is not included in these models, a lack that could result in an underestimate of growth rates, although the fitting procedure may have inadvertently considered this. The predicted growth rates under recommended flows (Hardy et al. 2006a, Figures 151-155) show generally larger fish than under existing conditions except at the end of the period modeled. The reasons for the decline at the end of the period are not explained but coincide with periods of predicted elevated temperature in the river. Hardy et al. (2006a) conclude that the recommended flows result in higher growth. With no information on the confidence limits of the predicted growth rates, one cannot judge whether the growth rates are significantly higher than observed under existing conditions. Similar concerns can be raised with respect to use of SALMOD to predict Chinook out-migrants (Hardy et al. 2006a, Figures 156-157). The recommended flows result in a higher maximum, a slightly higher mean, and a lower minimum number of out-migrants than other flow regimes. The significance of differences of this magnitude should be assessed.

Hardy et al. (2006a) use data and models from Dunsmoor and Huntington (2006) to show the effect of thermal conditions on upstream salmonid migration. Yet, the thermal conditions for the recommended flows during the critical autumn and spring migration periods were not analyzed, and how the migrants may respond was not discussed. Instead, a more general argument about larger flows resulting in larger spawning areas is provided. Another general argument is used with respect to out-migration, noting that the recommended flows move fish downstream more rapidly and provide greater access to cooler tributaries. These general arguments should be supported with data that facilitate comparing present conditions with the proposed instream flow recommendations.

Water Quality and Temperature

California has listed the Klamath River as impaired because of high temperatures, low dissolved oxygen, and excess nutrients. Hardy et al. (2006a, p. 47) state, "We believe that dissolved oxygen and other water quality parameters are of secondary importance to our efforts compared to that of temperature." This conclusion is based on water-quality simulations conducted as part of the PacifiCorp relicensing and the ongoing total maximum daily load (TMDL) for the Klamath River basin. Data should be presented to support this statement. In particular, low dissolved oxygen and elevated ammonia concentrations have been associated with disease outbreaks and fish kills in the river (Tetra Tech, Inc. 2007).

The impacts of temperature and water quality on river ecosystems are not limited to alterations of growth and survival of three anadromous salmonid species, although those are the only direct impacts included in models used by Hardy et al. (2006a). Temperature and water-quality regimes affect the species composition and productivity of aquatic primary producers, decomposers, invertebrates, and the other 16 native and 17 non-native fishes. These impacts on critical food-web components can indirectly alter salmonid growth and survival. They also influence the abundance of the parasitic myxosporean *Ceratomyxa shasta*, which has been associated with several juvenile fish kills in the river. A polychaete is the intermediate host for the parasite, and the polychaete is most abundant in areas of low flow and fine sediments where dense beds of the alga *Cladophora* occur (Stocking and Bartholomew 2004). Excess nutrients would stimulate the growth of these *Cladophora* beds, and hence could indirectly affect parasite abundances.

Temperature is modeled using two approaches, which appear to be adequate: SIAM, which uses MODSIM for flows and HEC 5Q for temperature and water quality; and the RMA modeling system, which uses RMA-2 (hydrodynamics) and RMA-11 (water quality and temperature). Diel temperature variation is potentially considerable and of biological concern (NRC 2004a), and both models estimate within-day variation. Hardy et al. (2006a, Figure 10) compared the mean daily temperature predictions of the two models but did not provide a comparison of observed and predicted temperatures; this is critical because predicted temperatures from RMA11 and SIAM are significantly different (almost 10° C in June based on Figure 11, Hardy et al. 2006a). Hardy et al. attribute this to a flow-induced bias but do not adequately explain the difference.

Hardy et al. (2006a) discuss a recent study (Tanaka et al. 2006) that showed hyporheic inflows that are coolest during times when the river is warmest and concluded these areas provide significant (up to 1,300 cfs) thermal refugia. The Tanaka et al. study examined thermal refugia provided by two tributary streams. Temperatures were always lower in the tributaries, but the refugia exhibited a diel pattern of refuge expansion and contraction with greatest temperature differences at night and early morning and least in the late afternoon. The size and constancy of thermal refugia were influenced by local geology, channel form, and the presence of algal beds. Subsurface seeps associated with tributaries, but at sites other than at the immediate confluence, provided some of the observed thermal refugia. The total number of fish in the thermal refugia studied increased when temperatures in the main stem were above 23° C. These data support the contention that tributaries can provide thermal refugia; however, the number, areal extent, and extent of overcrowding of fish in thermal refugia remain poorly understood.

Connectivity

The charge of the Hardy et al. (2006a) modeling effort was to examine the main stem of the Klamath River, in isolation from other geographic areas relevant to the fish life cycle, particularly the tributaries and the ocean. This is in contrast to the NRC (2004a) conclusion that any successful effort to restore anadromous fishes on the Klamath River is likely to require an emphasis on restoring ecosystem functioning in the tributaries, where most salmon in the basin originally spawned and reared. This is a fundamental limitation of the entire modeling effort.

Hardy et al. (2006a) have imposed greater hydrologic connectivity than other efforts in terms of estimating tributary inflows downstream of the dams to combine with main-stem inflows estimated at the dams. These tributary inflows are quite important and constitute the majority of the inflow for much of the stream for most of the year.

An overall problem with modeling efforts for this basin has been poor coordination and poor specification of purpose. Although the individual modeling efforts have been valuable in small but often important ways, the whole of the efforts is less than the sum of the parts. This is discussed more fully in the next chapter.

Scaling

Hardy et al. (2006a) did a good job of ensuring that the stream-habitat cell size determined by 2-D hydraulic modeling was compatible with fish-habitat use. Simulated habitat-modeling output for selected discharge levels was compared with independent field observations of fish distribution for several salmon life stages. This kind of testing is seldom done. The Hardy et al. study is a considerable improvement over most similar habitat studies. However, statistical analyses of the goodness of fit between model output and fish observations would provide a much-needed improvement.

Temporal scales were mixed among the models used in the overall analyses. The imposed monthly time step for the NFS placed a severe limitation on the Hardy et al. study. A more comprehensive hydrologic model of the stream network using a daily time step as input to the habitat and fish population modeling is recommended.

Integration

Hardy et al. (2006a) made a credible effort to integrate hydrodynamic and hydraulic habitat over the entire length of the river's main stem, given the resources and data available for the modeling effort. Hydrodynamic, water-quality, habitat, and population modeling are addressed for the main stem. However, the integration was somewhat piecemeal in that separate analyses were conducted of habitat, temperature, bioenergetics, and fish production. The flow recommendations were based on reconstructed habitat time series using flow-habitat relationships and NFS hydrology. These recommendations were then supported as being reasonable through temperature, bioenergetics, and population modeling after the fact. Refined analyses using daily time series of hydrology; habitat; water quality, including temperature, incoporated into the Chinook salmon population model SALMOD are needed. In addition, similar efforts for coho salmon, steelhead, and other species are warranted. If completed, then testing of movement, growth, and out-migration could be accomplished with monitoring in an adaptive-management context, as is being done on the Trinity River (Schluesner 2006). Water quality (dissolved oxygen, contaminants) and important processes such as sediment transport (accounting for gravel spawning beds) and channel formation (bar building, migration) need to be integrated.

The IFS focused only on information relevant to the main-stem Klamath River below Iron Gate Dam without any work in tributaries. The study team proposed to incorporate tributary systems, but members of the Klamath River Basin Fisheries Task Force opposed any instream flow assessment work in the tributaries (Hardy et al. 2006a, p. 49). Because tributaries were not considered, the usefulness of the model developed by Hardy et al. is more limited than it would be otherwise.

Monthly natural-flow estimates were used to generate stochastic flow series and then to generate habitat time series. A consequence of mean monthly flow series, which tend to average flood peaks and low flows, being used in the hydrodynamic model is that the habitat time series developed for the IFS may not represent flow variability at the most relevant time scale for biological processes. If the temporal scale of the habitat time series is not relevant for biological processes, then results derived from the habitat time series may be erroneous.

Quality Assurance, Quality Control, and Model Testing

Various forms of quality assurance, quality control, and model testing were conducted by Hardy et al. (2006a). These included the use of technical advisory groups, protocols and procedures for data collection, and various forms of model testing. Data quality-control efforts were particularly impressive. A large amount of hydrodynamic and fish-habitat-utilization field data was collected for several field sites.

Model testing can be conducted in several ways (Chapter 3). The following list identifies some common test methods and ways that Hardy et al. (2006a) tested model components:

Software tests.   By relying on fairly well-established computer codes, many software testing issues are avoided. Hardy et al. relied on previous studies by others for software testing the hydrodynamic models.

Numerical tests.   Hardy et al. relied on previous studies by others for numerical testing of the hydrodynamic models.

Empirical tests.   Hardy et al. collected an extraordinarily large amount of data on water depth, velocity, and habitat utilization for model development, calibration, and empirical testing. While large in quantity and high in quality, the study could afford to collect data only for seven reaches of the main-stem system. Although group consensus and agreement among interested parties is useful when selecting sample sites, in more contentious situations such as the Klamath River basin, additional statistical analyses are needed. Analyses comparing habitat sample reaches (and mesohabitats) within and among physiographic segments would provide useful support for the river stratification and representativeness of the sample reaches chosen.

Empirical testing of water-surface elevation model results was conducted for the reaches chosen to test the hydrodynamic component of the overall model. Extensive pictorial comparisons were made for the study-site reaches (Hardy et al. 2006a, Appendix E), but few statistical comparisons were made for water-surface elevations.

Limited comparisons were made of model results against field observations of local stream velocities (Table 5-2). Pictorial comparisons of field observations and model results for velocity data for several sites (Phase II, Appendix F) indicate broad qualitative agreement between model results and field velocities for four sites. A more formal statistical analysis of the ability of the model to predict field conditions would have been useful. This would be especially useful along the stream channel margins and overbank, as these are important habitat areas covered by simulations over the full range of flow conditions.

Empirical tests of habitat utilization also were conducted, albeit for a narrow range of flows, as noted by Hardy et al. (2006a). The absence of statistical analyses for these comparisons limits their usefulness. Field data were available for several species and habitat study sites (Hardy et al. 2006a, Figures 77-106). Habitat-model predictions coincided with available field observations for Chinook spawning and fry. For Chinook juveniles, there were relatively few observations, but those available coincide with favorable habitat predicted by the habitat model. For steelhead fry, the few field observations available matched the habitat model. For coho fry, no field observations were available, but model results matched observations of general habitat use reported by Stutzer et al. (2006). For coho juveniles, field data were too sparse, but habitats modeled as more suitable coincide with qualitative observations of local experts. These visual comparisons of

TABLE 5-2 Testing of Habitat-Suitability Model Results

Species	Spawning	Fry	Juveniles
Chinook	Sizable field-data comparison; good match	Sizable field-data comparison; good match	Few field data; good match
Steelhead	Not modeled	Few field data; good match	Sizable field-data comparison; moderate agreement
Coho	Not modeled	Results agree with observations by Stutzer et al. (2006)	Data too sparse for empirical test; results qualitatively agree with expert opinions

fish location versus modeled habitat suitability would be more defensible if statistical analyses involving fish presence-absence versus suitable-unsuitable tests were conducted as outlined by Thomas and Bovee (1993).

The numerous visual representations presented by Hardy et al. (2006a) to show the coincidence of fish observations and modeled habitat suitability illustrate that while fish observations usually occur in areas indicated as high suitability by the model, there are many high-suitability areas (indicated by the model) that have no observations. This raises the possibilities that the river was underseeded (well below carrying capacity for these fish species), that fish tend to cluster together for nonhabitat reasons, or that some of the model-predicted high-suitability habitat in fact is not desirable for fish. Although the pictorial comparison of model results and fish-observation data is useful and intuitive, a statistical test of the ability of the model to predict field observations is needed. The emphasis of the fish-observation study was primarily on testing the habitat-suitability criteria-derived usable-habitat output. Much more meaningful testing would involve the population dynamics of the fish species through space and time by monitoring and comparing fish movement, growth, and smolt out-migration with SALMOD simulations.

Sensitivity Analysis

Sensitivity analysis is difficult for a large integrated model with many hundreds of parameters. For hydrologic inputs, the stochastic hydrology developed by Hardy et al. provides sensitivity analysis. In addition, when coupled with a daily hydrologic model, the SALMOD model could be used to examine the sensitivity of the growth and movement aspects of the fish population model to changes in flow, physical habitat, and temperature (see Bartholow and Henriksen 2006).

IMPLICATIONS FOR IMPLEMENTING
FLOW RECOMMENDATIONS

Current Operations According to Water-Year Types

In 2004, the Klamath Basin Area Office of the USBR began issuing annual operations plans to provide estimates of water supplies to different areas served by the Klamath Project. The annual plan describes expected project operations from April 1 of one year through March 31 of the following year, based on current and projected hydrologic conditions within the basin (USBR 2006). Klamath Project facilities operate, to the extent possible, to provide lake or flow levels consistent with the biological opinions issued by USFWS and NMFS.

Initially, the annual plan is developed using April 1 forecasts supplied by the Natural Resources Conservation Service (NRCS) for the net inflow to Upper Klamath Lake. These forecasts are based on seasonal measurements of snow-water equivalent and precipitation at more than 30 sites within the region (Risley et al. 2005). The uncertainty in the April 1 forecasts for inflow to Upper Klamath Lake is about 20% of the long-term average flow, which is higher than in most basins in the western United States (Risley et al. 2005). Inflows to Upper Klamath Lake are estimated based on NRCS forecasts, with different criteria used for planning lake elevations and Klamath River flows (USBR 2006). Depending on hydrologic conditions, the forecasted water year is assigned to one of several categories. Four categories are used in lake-level planning (above average, below average, dry, and critical dry), and five categories are used for river flow planning (wet, above average, average, below average, and dry). As conditions change, the water-year type may be revised through June. The water-year type is finalized on September 30 (USGS 2005), and this designation remains in place until March 31 of the following year.

The different water-year categories for lake levels and river flows were established as a result of a series of discussions in 2002 between the USBR and USFWS (lake levels) and NMFS (river flows). In the Biological Assessment issued in February 2002, the USBR proposed to continue operating Klamath Project facilities consistent with hydrologic conditions observed in the 10-year period from 1990 through 1999; this period included a mix of the water-year types noted above (USBR 2002). Subsequently, NMFS issued a Biological Opinion raising concerns that the proposed Klamath Project operations would result in incremental depletions of flows below Iron Gate Dam (NMFS 2002). The Biological Opinion was based on information from several sources, including the Biological Assessment (2002), an interim report issued by the NRC (2002), published papers, and a draft report issued by Hardy and Addley (2001). The NMFS used unimpaired flow estimates provided in the Hardy-Addley report to recommend five water-year types: wet (10% exceedance), above average (30% exceedance), average (50% exceedance), below average (70% exceedance), and dry (90% exceedance) (NMFS 2002). The USBR adopted the five water-year categories in 2004, and these remain part of the annual operation plans. Similar categorization into five water-year types for river-flow planning is being applied on the Trinity River (www.trrp.net).

The use of only five water-year types, as specified in the current operational plans for flows at Iron Gate Dam, can present problems in meeting the flow requirements in years when there is a change in the water-year type resulting from a change in hydrologic conditions, for example, from below average to dry (USGS 2005). The potential problem is compounded by the time steps used in adjusting flows at Iron Gate Dam (2 or 4 weeks, depending on the time of year). The USGS has begun investigating the feasibility of using a water bank to augment flows to assist in meeting the instream flow requirements at different times of the year (USGS 2005). Increasing the number of water-year types would result in smoother shifts in flow patterns if the designation were changed. However, there are practical limits to the number of categories that can be used to set flow requirements. Bigger concerns in evaluating present operations and flow requirements are associated with the uncertainty in the NRCS forecasts of inflows to Upper Klamath Lake, as noted above, and the representativeness of the period of record (1990-1999) used to develop the flow requirements in the 2002 Biological Opinion. The USGS studies under way indicate that the 10-year reference period results in abrupt monthly shifts in the required flows due to the relatively small number of years in each water-year type. Furthermore, their analysis suggests that the reference period, 1990-1999, is not representative of the current hydrologic conditions in the system. The USGS proposed alternative is to set flow levels below Iron Gate Dam based on data and information from the 39-year period of record (1961-1999), as this would result in more hydrologically realistic requirements (USGS 2005, p. 35).

Instream Flows Proposed in the Phase II Report

In the Phase II report, the instream flow recommendations for the Klamath River below Iron Gate Dam are listed as a series of monthly flows corresponding to annual exceedance levels ranging from 5% to 95% in increments of 5% (Hardy et al. 2006a, Table 27). The different flow levels are not defined by water-year types (wet, average, dry), and the rationale for assigning such a large number of flow categories (19) is not stated. From an operations standpoint it may be impractical to implement the flow recommendations at this level of detail. Furthermore, the uncertainties in NRCS estimates of water supply in the upper basin are likely to be much greater than the differences in 5% exceedance levels listed, making it difficult to know in advance precisely where in Table 27 one would begin adjusting project operations to implement the flow recommendations. By using an extended period of hydrologic records as recommended by USGS, an increased number of water-year types to perhaps 10 of the exceedance levels, as is presented in the Phase II report, would smooth shifts in flow patterns.

As noted elsewhere in this chapter, it was difficult to follow the steps used to arrive at the monthly flow values listed in Hardy et al. (2006a), Table 27. These values apparently represent an average of monthly flows developed from considerations of the "natural flow paradigm" and the physical habitat provided for individual species through time series simulations derived from the NMFS (2002, p. 180). To place these recommendations in the context of previous recommendations and present operating requirements, Figure 5-11 shows a comparison of the monthly flow values listed in Table 27 of Hardy et al. (2006a) with monthly values listed in USBR's Biological Assessment (USBR 2002, Table 5.9) and the Klamath Project's operations plans. The colored lines represent flows corresponding to each of the annual exceedance values ranging from 5% to 95%, and the colored bars represent the five water-year types. This figure shows that, except for "wet" water years, the instream flows recommended generally exceed the flows proposed in the 2002 Biological Assessment. Figure 5-12 shows a similar comparison between the flow recommendations listed in the IFS and the long-term (2006-2012) flow recommendations given in the NMFS Biological Opinion. Except for "below average" and "dry" water-year types for the months of May and June, all other long-term flows are lower than Hardy et al.'s instream flow recommendations.

The instream flow recommendations of Hardy et al. (2006a), based on the natural-flow paradigm, were mostly derived from the naturalized flows predicted at the Iron Gate Dam. Implementation of the recommended instream flows using annual exceedances (Hardy et al. 2006, Table 26) ideally requires the prediction of the expected "hydrologic regime" under natural conditions. Current forecasts made by NRCS are likely based on methods developed using recent flows, which have been affected significantly by agricultural practices in the upper Klamath River basin, into Upper Klamath Lake. Instead, the forecasts of "annual exceedance type" as opposed to the "water-year type" in the current implementation should be based on predicted flows under natural conditions. Under such conditions, precipitation (or snow pack) should be the primary hydrologic variable that influences the prediction of the natural-flow regime in a given year. It appears that the precipitation during the October through March period has a good correlation to the naturalized flows during the April through September period (Figure 5-13).

Consequently, precipitation during the first 6 months of the water year may be used to determine the "annual exceedance type" of the following operational period, and the corresponding instream flow recommendations may be used for operational purposes. The precipitation class intervals corresponding to annual exceedances of the naturalized flows can be determined from the precipitation–natural-flow relationship (see Figure 5-13). Alternatively, the current NRCS method could be altered to predict the naturalized flows for selecting the "annual exceedance type" under natural conditions. Further, if a "natural" watershed model were available for the Klamath basin, the naturalized flows could be forecasted using the observed precipitation in the basin, and such forecasts could be used to update the annual exceedance types through summer and fall months.

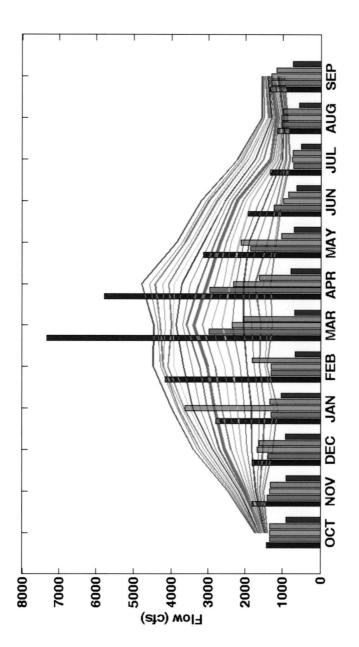

FIGURE 5-11 Comparison of monthly flows for Klamath River below Iron Gate Dam based on the NFS. Bars represent flows for each of five water-year types, as specified in Table 5.9 (USBR 2002). Lines represent flows corresponding to each of annual exceedance values ranging from 5% to 95%, as specified in Hardy et al. (2006a), Table 27.
SOURCE: Hardy et al. 2006a. Reprinted with permission; copyright 2006, Utah State University.

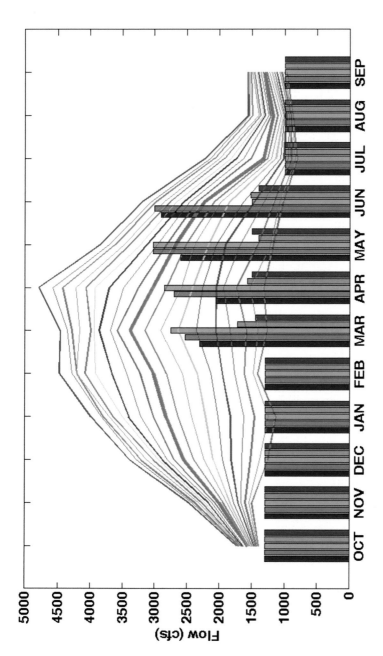

FIGURE 5-12 Comparison of modified Table 9 flows (NMFS 2002, USGS 2005) with Hardy et al. (2006a) instream flow recommendations. Bars represent modified Table 9 flows, as specified for each water-year type. Lines represent recommended instream flows corresponding to annual exceedance values from 5% to 95%.
SOURCE: Hardy et al. 2006a. Reprinted with permission; copyright 2006, Utah State University.

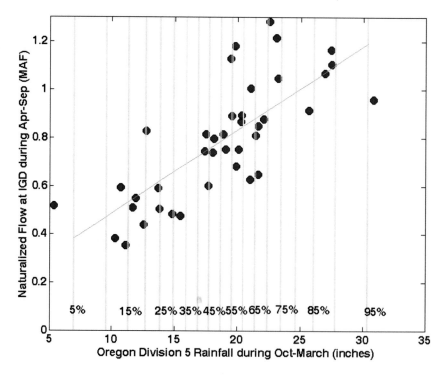

FIGURE 5-13 Relationship between October-March precipitation and naturalized flows at Iron Gate Dam during April-September period. Vertical lines correspond to annual exceedance levels for precipitation derived from the equivalent naturalized flow values.

COMPREHENSIVE ANALYSES AND INTEGRATION

Hardy et al. (2006a) provide several important initial steps (including novel and valuable data) toward a comprehensive Klamath River basin management program. The temperature and hydraulic modeling components have been accomplished to the degree necessary for the main-stem Klamath River. Similar efforts are ongoing for the Trinity River. However, tributaries to the Klamath and Trinity rivers, which are important to the life history of coho salmon, have not been quantitatively addressed. As is true of all instream flow studies, a comprehensive hydrologic description of the river system (including tributaries) is fundamental to system understanding and management. The existing hydrologic data are not sufficient for the level of analysis that is necessary. An understanding of the relations among temporal and spatial variations in hydrologic conditions and the resulting habitat, temperature, and fish-population dynamics is essential for developing management prescriptions leading to sustained recovery of the anadromous species.

Adequate management of Klamath River salmonids requires a network-level hydrologic model capable of generating daily time series of stream discharge at numerous points throughout the river system (including tributaries). This level of modeling is essential for using the spatial and temporal dynamics of the temperature regime and habitat modeling to elucidate possible "habitat bottlenecks" and as input to appropriate fish population-dynamics models (species-specific SALMOD). Comparing "synthesized natural and existing" hydrologic conditions and subsequent development of instream flow prescriptions for recovery of the salmonid species is only appropriate if the various models use the same time step. Such information is critical for recommending instream flows for a Klamath River basin water-management program. Unfortunately, there is little or no synergy between the two reports the committee was charged to review. The NFS presents monthly data below Link River Dam, whereas the temperature model used by Hardy et al. (2006a) provides daily data down the main-stem Klamath River, and the habitat and fish-population analyses used a monthly time step and SIAM model data.

A second point of concern is the apparent lack of emphasis on the sediment dynamics of the system. Changes in management of a river system require study of the sediment input (or lack thereof) and sediment transport and discharge relations. Any river system with dams and altered discharges will have changes in the sediment balance and perhaps alterations in the channel form and sediment distribution. Valuable data from such studies could address the present state of the river channel (aggrading, degrading, or in some stage of dynamic equilibrium) and better inform management decisions on maintaining the existing channel, if in a state of dynamic equilibrium, or driving the system toward a new equilibrium state. Hardy et al. (2006a) do not discuss this issue, although reference to the Ayers report (Ayers Associates 1999) implies an assumption that the present channel is not actively aggrading or degrading. The roles of overbank flows (channel forming, recruitment of nutrients, and fish refuge) and sediment-flushing flows are important components of a comprehensive IFS and should result from sediment-transport and channel-dynamics studies. Hardy et al. (2006a) assumed that these flows would be present, but the necessary flows for maintaining channel dynamics were not quantified. Without such quantification, there may be no protection from future flow depletions resulting from diversion and storage of peak flows.

Another apparent impediment to the integration of science and management strategies is the jurisdictional separation of the Klamath River and Trinity River main stems into two separate USBR Area Management Offices. A thorough understanding of salmon stocks native to the whole Klamath River system and of how management of the water supply may help to achieve their recovery will require system-wide analyses and joint management. Following the mass mortality of adult salmon in the lower Klamath River in September 2002, there has been increased pressure to integrate efforts of the Trinity River Restoration and Klamath River Task Force. For example, summer releases from the Trinity River could provide much-needed cooler water in the lower Klamath River, but if done routinely each year, summer releases might conflict with Trinity River adaptive-management objectives for attaining appropriate variability within inter-annual flow regimes. Integration of these two efforts is essential. The Trinity River Restoration Program is in the process of collecting habitat, temperature, and fish-population data similar to those presented by Hardy et al. (2006a), and a data-management system is being developed for use by both the Klamath River and Trinity River programs (R. Wittler, USBR, personal communication 2006). To fully integrate these two programs, the same models (such as flow, temperature, habitat, and fish-population dynamics) with the same level of detail, time step, and linkage among models should be used on both sub-systems. With similar integrated modeling and data management, alternative management scenarios (involving reservoir releases, sediment augmentation, and so forth), using an assumed water supply, climate conditions, and salmon stock returns, could be quickly evaluated. Such linkage of models could greatly facilitate communication among various stakeholders and managers and lead to better adaptive-management approaches. Integration of these efforts with ongoing activities in the Shasta and Salmon rivers also would be beneficial.

CONCLUSIONS AND RECOMMENDATIONS

Having reviewed the IFS (Hardy et al. 2006a) the committee finds that it enhances understanding of the Klamath River basin ecosystem and the flows required to sustain it. However, the flow recommendations were based on manipulations of the PHABSIM modeling and supplemented by existing flow records in ways that did not clearly derive from any theoretical considerations. The steps used to derive the final flow recommendations were not pre-specified or thoroughly tested.

In its present form, to the degree they are adopted, the recommended flows resulting from the study should be adopted on an interim basis pending the model improvements outlined below and a more integrated assessment of the scientific needs of the basin as a whole. The recommended flow regimes offer improvements over existing monthly flows in that they include intra- and inter-annual variations and appear likely to enhance Chinook salmon growth and young-of-the-year production. More detailed discussion of the study's implications for the basin's fishes is in the final section of this chapter.

In this report, the committee is critical of many aspects of the IFS, but the committee also found substantial strengths in the study. The following paragraphs outline the substantial strengths and contributions of the IFS that managers and investigators may find useful for decision making.

The IFS represented a state-of-the-art process for the modeling of temperature and bioenergetics for riverine fish species. Temperature modeling is especially important for fishes in the lower Klamath River because temperature may be a limiting factor in late summer for suitable habitat. Understanding how temperature varies, particularly in response to flows that likely result from dam releases, may provide support for more effective operating rules during low-flow periods. High temperatures affect the survival of fishes directly if they are high enough to kill the fish; they also can affect them indirectly by affecting bioenergetic processes. A prominent feature of the study is the Chinook salmon fry and juvenile growth modeling using the Wisconsin Bioenergetics Model (Stewart and Ibarra 1991, Hanson et al. 1997). Bioenergetic analysis of fish growth often is not included in instream flow assessments.

The IFS is broadly consistent with the guidelines of Instream Incremental Flow Methodology (IFIM). While the committee has reservations about the use of some models (including PHABSIM) that support the connection between flow and habitats, IFIM is widely used as a general approach to connecting research associated with flows to management decisions. Generally speaking, the IFIM process includes legal and institutional analysis, strategy design, technical scoping, habitat modeling, definition of alternatives, feasibility studies, and negotiated resolutions with stakeholders (USGS 2007b). The IFS for the Klamath River fits well into this approach in the habitat modeling portion of the overall process.

Incorporation of salmon fry behavior and distance to escape cover is an exceptional advancement offered by the IFS. Salmon fry are a critical part of the salmon life cycle in the river environment, and their inclusion in the model provides increased confidence in the model predictions. By examining the micro-geography of the stream channel, the IFS provides the ability to examine how the distribution of fishes interacts with the distribution of potential protective cover for them, a metric that is substantially variable over short lengths of river channel. Many instream flow studies do not assess this distance to cover because of the stringent data requirements, but the Klamath study successfully addressed this concern.

The Klamath instream flow model makes a substantial contribution to decision making with its comparison of consequences of implemented flows with existing flows for smolt production, a critical step in the overall life cycle of the anadromous salmonid fishes. Most instream flow models create a set of predictions and stop at that point, but the Klamath study goes one important increment further by comparing the existing flow regime, which might be suspected of inhibiting smolt production, and drawing quantitative comparisons with a proposed flow regime that might improve smolt production. The result provides managers with useful insights into factors that are important for decision making and provides scientific insights to the general salmon survival process.

In another comparison between predictions and observations, the IFS assesses predicted and observed fish locations in representative reaches of the river. Generally speaking, there was considerable agreement among predictions of fish distributions through habitats in the reaches that were examined. This agreement lends some credibility to the model, and further comparisons can show under what river conditions the model is strongest and under what conditions it is weakest. The committee finds that many similar studies fail to test their output, as was done in this case.

Finally, an important strength of the IFS is its state-of-the-art application of flow models in simulation of habitat suitability. Flow models have gradually evolved over the past several decades from one-dimensional representations with assessments of flow variation only in the downstream direction. While useful, such assessments did not reflect the enormous complexity of river channels, particularly relatively large channels, such as the Klamath River. Recently, hydraulic researchers and engineers have made increasing use of two-dimensional flow models that can trace variation in the cross-channel direction in addition to the downstream direction. Few instream flow studies have yet to take full advantage of the newer two-dimensional models, partly because of their demand for large amounts of data describing the topography of the bed and banks of the channel as well as nearby riparian areas and flood plains. Two-dimensional models also require extensive computing capabilities. The IFS has made good use of new techniques and extended computing capability to improve the understanding of the fluvial complexities of salmon habitat in the river.

Specific conclusions and recommendations follow.

Conclusion 5-1. The goal of the IFS is to recommend "flow regimes that will provide for the long-term protection, enhancement, and recovery of the aquatic resources within the main stem Klamath River." This study was limited because achieving that goal requires daily flow data and the USBR NFS provided monthly flow data.

Recommendation 5-1. The IFS should be updated using daily flow data. The monthly data could be disaggregated into daily data before hydrologic time series are used in habitat modeling, or a daily watershed model that represents each of the major sub-basins of the Klamath River basin under natural conditions could be developed. Such forecasts could be used to update the annual exceedance types through the summer and fall months. Future investigations could compare the unregulated flow regime of the river with present conditions and seek to restore as much of the functional river-based ecosystem as possible through managed flows benefiting a full complement of species.

Conclusion 5-2. Although the methods used for habitat assessment in the IFS extend the standard practice for assessing habitat-flow relations by incorporating escape cover and state-of-the-art hydraulic modeling, there are limitations in assessing primarily the hydraulic aspects of habitat, independent of other river-ecosystem attributes (for example, physical processes, temperature, water quality, bioenergetics, and fish production). Therefore, more integration of the analytical models is needed. Better-integrated models also would facilitate adaptive management of the river system.

Recommendation 5-2. Habitat modeling should integrate hydraulic analyses with geomorphic processes of sediment transport (for example, sediment-flushing and channel-forming flows), water quality (for example, temperature, dissolved oxygen, and contaminants), and fish-population dynamics. In addition, an adaptive management strategy should provide a clear context of management decisions to which habitat modeling results would be relevant.

Conclusion 5-3. In addition to the integration recommended above, successful maintenance of aquatic resources in the Klamath River depends on several aspects of water quality in addition to temperature conditions, including dissolved-oxygen levels, nutrient concentrations, sediment loads, and contaminants. The IFS does not present analyses of these water-quality attributes, nor does it justify excluding them from analyses. Failure to analyze the thermal conditions for the recommended flows during the critical autumn and spring migrations is another shortcoming of the study.

Recommendation 5-3. An addendum to Hardy et al. (2006a) should be prepared that includes analyses of several aspects of water quality, such as dissolved-oxygen levels, nutrient concentrations, sediment loads, and contaminants, or the addendum should justify excluding these important considerations.

Conclusion 5-4. Tributaries of the Klamath River are important for maintaining anadromous fish populations because they provide essential habitat and are sources of water and sediment that maintain main-stem habitats.

Furthermore, although coho salmon and steelhead are found in the main stem, tributaries contain the most important habitat for producing juveniles of these species. Since technical assessments conducted as part of the IFS were confined to the main-stem Klamath, the usefulness of the study for evaluating coho salmon and steelhead management options is severely limited.

Recommendation 5-4. Future studies should include explicit analyses of the habitat, water, and sediment contributions of tributary streams in the context of the fish life histories (particularly coho and steelhead) and movements throughout the entire Klamath River basin. They should assess the ability of tributaries to facilitate juvenile fish production and thermal refugia and provide estimates of tributary habitat, their areal extent, and the extent of overcrowding of fish in them.

Conclusion 5-5. The IFS lacks adequate statistical testing of model predictions. Specifically needed are statistical analyses of the comparison of observed fish distributions with predicted distributions of usable habitat as defined by simulated hydraulic and temperature conditions, confidence limits on the fish-growth predictions, and statistical testing of hydraulic-model velocity predictions for the entire range of flow conditions in the channel and overbank. These analyses would increase confidence in the validity of modeling results and conclusions.

Recommendation 5-5. An addendum to Hardy et al. (2006a) should be prepared that contains results from rigorous statistical testing of model outcomes, including a comparison of observed fish distributions, predicted temperatures, and confidence limits on growth predictions. Just as habitat simulations were made for the entire range of flow conditions, statistical analyses should be presented to support velocity predictions in the channel and overbank.

Conclusion 5-6. The approach used in the IFS apparently assumes that physical habitat is an important limiting factor to recovery of the salmonid fishes. However, the study does not demonstrate when (or if) habitat may be limiting to the fish species and the identification of potential life-stage "bottlenecks" when comparing existing and naturalized flow time-series simulations. Habitat-duration curves alone are not sufficient to illustrate which life stages may be most vulnerable. SALMOD modeling would be more appropriately used for analyses of potential habitat limitations imposed by hydraulic and temperature conditions, sensitivity analyses, and comparison of salmon-smolt growth and production between naturalized and existing flow regimes.

Recommendation 5-6. Investigators should provide a thorough analysis of the life stages of salmonid species, allowing for comparisons of seasonal differences in usable habitats between naturalized and existing flow simulations and for identifying possible habitat limitations imposed by present conditions and potential improvements provided through recommended flow regimes. Full integration of hydrology, habitat, temperature, and Chinook salmon life history with SALMOD is needed. SALMOD modeling should be used for identifying habitat limitations and developing alternative instream flow recommendations, not simply for post hoc testing. Testing of salmon movement, growth, and production through ongoing monitoring efforts should accompany these modeling efforts, as is being done on the Trinity River. Such integrated and tested modeling capabilities could prove useful for future adaptive-management efforts.

Conclusion 5-7. Given the overall objectives of the IFS, a reasonable process and rationale were used to stratify the main stem Klamath River into five "homogeneous" study reaches and to identify seven representative study sites. However, the representativeness of the study sites was determined by inter-agency group agreement and was not statistically assessed. If the mesohabitat distribution within the study sites is not shown to be representative of the larger study reaches, there will remain significant uncertainty about the efficacy of the node-weighting approach used to transform (that is, upscale) site-specific habitat-suitability criteria for the study reaches.

Recommendation 5-7. The IFS should include an analysis that demonstrates that the study sites are representative of the respective study reaches by providing field measurements of basic channel properties at intervening locations among the study sites. In addition, analyses are needed to demonstrate the efficacy of upscaling site-level habitat modeling results to the study reaches.

Conclusion 5-8. Hardy et al. (2006a) used a PARMA model to generate a set of naturalized flows for the Klamath River. There are three potential limitations in their approach. First, the PARMA model may not permit an accurate assessment of uncertainties in the naturalized flows simulated using the water-budget approach. Second, the model may not adequately capture serial (intra-annual or inter-annual) autocorrelation and spatial correlation in the data. Finally, investigators may have fitted the model without properly accounting for skewness in monthly flow distributions.

Recommendation 5-8. The analyses required to address limitations to the PARMA model developed by Hardy et al. (2006a) should be conducted, and a formal assessment of model inaccuracies should be conducted.

Conclusion 5-9. The IFS identifies five necessary instream flow components: overbank flows, high-flow pulses, base flows, subsistence flows, and ecological base flows. Among these, the discussion of overbank flows states that operation of the existing Klamath system provides large discharges during wet periods that meet or exceed the channel-maintenance requirements, but there is no discussion of the importance of overbank flows, particularly their frequency and duration. In addition, the study defers discussion of high-flow pulses (citing ongoing testing) and assumes that the maintenance of intra- and inter-annual habitat values at or near "natural flow" accounts for base, subsistence, and ecological base flows.

Recommendation 5-9. Specific high-flow recommendations for maintaining channel integrity are needed. Additional physical-process studies are also needed to address sediment transport and establish threshold levels of discharge that maintain channel dynamics. Results of these studies need to be incorporated into variable instream flow recommendations.

MANAGEMENT IMPLICATIONS OF THE INSTREAM FLOW STUDY

The basic conclusions of the IFS are recommended flows expressed as monthly target values for discharges below Iron Gate Dam on the Klamath River. The study adopted the Natural Flow Paradigm, and its primary input was the natural flows defined by the NFS. The IFS integrated the NFS flows and the historical flow records with water-temperature simulations to accommodate existing understanding that flow volume and water temperature are two primary controls on fish growth and survival. The most important outcome of the IFS was that it indicated that increases in existing flows downstream from Iron Gate Dam probably would benefit fish populations through improved physical habitat associated with more water and through reduced water temperatures. The increased flows would reduce slack-water areas that are disease-prone areas for salmon. Tests conducted in the study led the authors to conclude that if the prescribed flows had defined the river's hydrology instead of the actual regulated flows (mostly less than the prescribed flows) during the period 1949 to 2000, salmon production in the lower river would have been higher than it was under the prescribed flows.

If these conclusions were borne out by studies incorporating experimental flows and monitored responses, managers would be able to have greater confidence that decisions to increase flows would be likely to have a beneficial effect on anadromous fishes in the lower river. The authors of the IFS mention two caveats, and the committee agrees with them. First, the flow recommendations apply to the needs of the anadromous fishes in the lower

Klamath River, and they do not account for competing water demands for other purposes, such as agricultural needs or the needs of federally listed fishes in the upper basin. Second, the flow recommendations address the needs of all the anadromous species in the lower Klamath River. They are not targeted for any individual species (listed or otherwise), and it is not possible to evaluate the conclusions separately for individual species.

The committee has additional caveats about the study's results. The flow recommendations are limited to monthly values because of the nature of the input from the NFS. This characteristic implies for users that the recommendations are general in nature and that they are useful for general planning. Because there are no daily flow recommendations, and because of various limitations on the calculations outlined in Chapter 5, the recommended instream flows do not provide guidance for daily discharge rules for system operations. The recommended instream flows are applicable to management of all anadromous fishes in the lower Klamath River, and there is no specification by species (Hardy et al. 2006a, p. 215). Of at least equal importance, the study, like the NFS, does not address the Klamath River's tributaries. Those tributaries are of great importance to the hydrologic regime in the Klamath River, especially below the confluence with the largest tributary, the Trinity River, but also above that point. To the degree that the anadromous fishes use the tributaries for spawning and rearing habitat, they are important to the productivity of those species. Finally, as the committee has detailed in Chapter 5, there are sufficient uncertainties and flaws associated with the study to show that it cannot be used as a specific guide to specific flows with much confidence.

Despite all the foregoing, it is extremely unlikely, in the committee's judgment, that following the prescribed flows of the IFS Phase II would have adverse effects on any of the anadromous fish species. Based on general principles and the information developed in that study, following its prescribed flows probably would have some beneficial effects on the suite of anadromous fishes in the Klamath River considered as a whole, although not necessarily for every species.

The conclusions and recommendations appearing in this and the preceding chapter lead the committee to consider the NFS and the IFS of the Klamath River in a more general context. The river, its waters, species, and habitats along with the myriad of human-induced controls operate in a complex ecosystem defined geographically by the watershed. The next chapter considers these larger-scale issues as a way to better understand the validity of the science and engineering approaches to the NFS and IFS and provides a framework for integrating resource management, science, and stakeholder concerns.

6

Applying Science to Management

INTRODUCTION

The weaknesses in the models addressed in previous chapters largely result from an ad hoc development and application of science to ecosystem management in the Klamath basin. Those weaknesses also suggest a lack of consensus regarding which hydrologic and biological facts are most pertinent to Klamath River operations and their effects on agriculture, the river's salmon fishery, and imperiled species. Similar concerns were expressed in the previous National Research Council (NRC) report on the Klamath River fishes (NRC 2004a), which described ecosystem management in the Klamath basin as "disjointed, occasionally dysfunctional, and commonly adversarial." Disappointingly, nearly 4 years later those descriptors still appear to apply. The previous report explicitly identified key gaps in knowledge regarding Klamath basin fishes, and stated that implementation of the Endangered Species Act "cannot succeed without aggressive pursuit of adaptive-management principles, which in turn require continuity, master planning, flexibility, and conscientious evaluation of the outcomes of management." Applying an adaptive-management framework to ecosystem management remains essential and notable by its absence.

The utility of the ecosystem models assessed in this present report, no matter how scientifically sound the models might be, is in their application in managing flows, native fishes, and the Klamath River ecosystem. These applications are not possible without institutional links, or at least communication, among those who develop and run the models, those who make decisions about water allocation and flows, and habitat management.

Model products and the data they are based on should serve to inform the dialogue among policy makers, management agencies, scientists, and the basin's water users over long periods of time and for specific decisions. Individual models, no matter how sophisticated or complex, should not be viewed as having by themselves the potential to solve complicated problems of ecosystem restoration.

Models are intended to improve conceptual understanding and transparency of ecosystem management problems and solutions and to provide the ability to experiment with and test potential solutions with less field experimentation, time, cost, and ecosystem risk. The essential role of science in ecosystem management is generally acknowledged, as is the method of delivering science to large-scale resource management efforts, referred to as adaptive ecosystem management, or adaptive management. Unfortunately the Klamath River basin currently lacks institutional structures and relationships or decision-making arrangements that can facilitate adaptive management.

The committee does not presume to know the exact contours of a mechanism for dealing with the intersection of science and policy for the Klamath River basin, because these arrangements are best designed by the people who live there and who participate in the agency frameworks already in place. However, the committee can provide general guidance in the form of identifying the various attributes that such arrangements should have. The committee drew upon experiences from other parts of the United States to learn what these attributes might be and how they might fit together in a coherent structure for science and policy. The following pages outline these attributes of successful management mechanisms, focusing on the application of adaptive management to the Klamath River.

The committee's conclusions on these topics, in addition to agreeing with and following up an earlier NRC (2004a) report, also largely share the spirit of the recommendations of at least two other reports on the basin, that by the so-called OSU-UC Davis Group (Braunworth et al. 2002) and the report of the Independent Multidisciplinary Science Team (IMST 2003). Perhaps the agreement among such diverse groups would lead to optimism for the future.

ADAPTIVE MANAGEMENT

Adaptive management is "learning by doing," wherein management actions, system modeling, data gathering, and decision making interact to maximize information gains, allowing for increasing management effectiveness and efficiency (Holling 1978). The details and scope of adaptive-management programs are diverse in type and application, but they frequently are categorized in one of three ways: efforts involving trial and error not based on an underlying hypothesis or conceptual ecosystem model, active adaptive management, or passive adaptive management. This committee does not endorse hypothesis-free trial and error approaches, because they inevitably are less effective and efficient than other hypothesis-based versions of adaptive management, and they frequently are more costly (NRC 2003, 2007b).

The most powerful form of adaptive management is the active form, which uses available information to develop models of how the system might respond to various decisions and conditions and allows the assumptions of the models and their conclusions to be tested experimentally in the field (see NRC 2007b). The combination of modeling capability and the experimental approach to management helps to identify the benefits and disadvantages of various management choices. Some managers are reluctant to be seen as experimenting with public resources, but in practice, choosing management options without experimenting is even less conservative.

A more-widely used approach is passive adaptive management, which is based on available information to construct "best-guess" conceptual models of the managed system, with management choices informed by a generally accepted conceptual model. Passive adaptive management typically is based on a single alternative developed from a conceptual model followed by monitoring and adjustment, whereas active adaptive management is based on alternative hypotheses that are examined experimentally. Monitoring and adjustment are critical components of any adaptive-management approach. Science in adaptive management responds to clearly articulated management needs for information.

Many restoration programs use a cycle of planning, management action, and data gathering, but the approach described in the literature of the CALFED Bay-Delta restoration program is particularly useful, as it shows a relationship between research and monitoring in adaptive management as they play sometimes distinct, but often complementary, roles in meeting the comprehensive restoration goals (Figure 6-1). It is too early to know how well the approach has succeeded (or will succeed) in practice.

Three critical attributes of adaptive management link science to management and policy development; hence adaptive management requires a programmatic (organizational) structure that facilitates communication and cooperation in decision making. Adaptive management requires the following:

(1) Clear articulation of program goals, with a description of anticipated application of information derived from research, monitoring, modeling, and risk analysis.

Adaptive-management programs need to determine whether current or proposed management practices or actions are maintaining the target en-

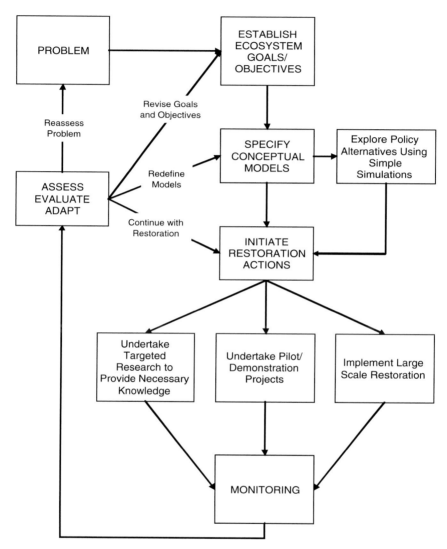

FIGURE 6-1 Flow diagram for adaptive management of scientific activities.

vironmental system and the ability of the system to deliver expected goods and services (examples in the Klamath basin would include numbers of salmon smolts or erosion control by vegetation). No universal set of goals or objectives can characterize a "high-quality" environmental state or can apply to all ecosystems subject to management and monitoring. But each proposed or current management action for which monitoring is intended should be accompanied by a set of specific project goals that are used to guide the development of monitoring objectives. The Comprehensive Everglades Restoration Plan (CERP) is a good example of identifying goals and performance measures for a restoration plan (NRC 2007b).

Management goals take many forms: they can be articulated in reference to a legal mandate (for example, associated with recovery goals under the Endangered Species Act or as attainment goals under the Clean Water Act); they can depend on science that has informed legislation (for example, the mandate in the Florida Forever Act to reduce phosphorus concentrations in the Everglades); or they can emerge from scientific studies that have been properly framed to help address management issues (for example, fishing restrictions pursuant to the Magnuson-Stevens Fishery Conservation and Management Act based on properly conducted stock assessments, and the Trinity River Restoration Program goals, which were informed by data from the Trinity River Flow Evaluation Report). Whatever the basis for a management goal, it should be articulated so that clear, quantifiable objectives can be identified and direct the monitoring design, and it should be based on the best available information vetted through program participants with policy, management, and scientific expertise, and with input from stakeholders.

(2) Development of models of the system to be managed that describe ecosystem attributes and the environmental stressors that perturb them.

The quantitative ecosystem models reviewed in this report show substantial sophistication and are highly evolved reflections of specific understanding about the hydrology and certain aspects of the ecology of the main-stem river. Such models often emerge from conceptual models (see Chapter 3) that have been vetted in their much simpler form and have benefited from review and some degree of general acceptance. Well-designed, generally agreed upon conceptual models enable research and monitoring programs to investigate relationships between causes of environmental perturbations and likely consequences. Conceptual models outline the interconnections among ecosystem elements and environmental stressors, the strength and direction of these links, and attributes of the system that can be used to characterize the state of resources. Conceptual models should include a representation of how environmental systems work and should emphasize anticipated responses to natural and human-caused disturbances.

Conceptual models should explicitly link ecosystem attributes, which include both abiotic and biotic elements and inputs, to system stressors. The expected cause-and-effect relationships that result in ecosystem changes in the model should guide the selection of candidate indicators for measurement in the monitoring program. Vetting conceptual models with program participants can document any consensus about facts pertaining to the resource management challenge, as well as technical or scientific controversies in particular need of attention.

There are inevitable barriers to the attainment of management goals and the success of restoration efforts. These barriers arise from the actions of both human-generated and natural environmental "stressors." Stressors are physical, chemical, or biological entities or phenomena that harm ecosystems and their constituent elements; stressors include wildfires, exotic species invasions, stream diversions, changes in stream temperatures, and conversion of natural vegetation to agriculture. Disturbances or stressors can be categorized for monitoring-plan development based on such attributes as characteristic frequencies of occurrence, extent of occurrence, magnitude (in both intensity and duration), selectivity (elements of the system that they act on), and variability. Disturbances or stressors that act on managed ecosystems need to be described in terms of causes and effects. That causal description is best presented as a conceptual model that links environmental stressors to environmental attributes of concern. Discussions among adaptive-management participants regarding the roles, intensities, and interactions of system stressors can serve as a forum in which differences of understanding and opinion about how the managed ecosystem operates can surface. Where greater knowledge exists, partial or complete quantitative models can be developed. Quantitative models can provide greater precision for developing promising solutions and experiments and are commonly more testable in the field than conceptual models.

(3) Selection of representative indicators of ecosystem status.

Ecosystems are far too complex to permit measurement of all of their attributes. Therefore, ecosystem conditions, their responses to restoration actions, and their susceptibility to long-term change must be assessed using a limited set of indicators. The theory and practice of indicator selection is demanding (for example, NRC 2000, 2003, 2005c, 2007b); the selection of the "wrong" (ineffective) indicators can cause a monitoring program to fail, as can the selection of too many indicators to be monitored with available resources. In addition, indicators that show intuitive or demonstrated relevance to program objectives can contribute to public support for the science and restoration efforts.

The most effective indicators possess several key attributes—they respond similarly to the dynamics of the greater ecosystem of concern; they respond rapidly to changes in their environment; their changes in status can be accurately measured; their natural variability is sufficiently limited such that changes in response to management can be differentiated from background variation; and they can be measured cost-effectively.

For monitoring in support of Klamath operations, at least three categories of monitoring indicators can be recognized (see also NRC 2000). First are function or process indicators, which measure ecosystem processes and their rates. Processes might include primary productivity, nutrient cycling, sediment accumulation, or water flows. Second are indicators of ecosystem structure, which can be used to assess ecosystem structure at any spatial scale from basin-wide distributions of spawning gravels to riparian vegetation distributions and connectivity along a tributary reach. Third are species-based indicators, which are important for the environmental-restoration program with its focus on at-risk and listed species. Species may be selected as indicators because they are members of groups thought to be important to ecosystem functioning (predators, primary producers, decomposers), may provide some insights into the functioning of the ecosystem (that is, they may serve as umbrella or keystone species or may be ecological "engineers"), may be the direct targets of management (because they are recognized as threatened or endangered), or may be especially sensitive to ecosystem change. It is striking that even the better-studied anadromous fish species, Chinook and coho salmon and steelhead, are not nearly as well understood in the Klamath basin as they could be; other anadromous species that also are important, such as Pacific lamprey, green and white sturgeon, eulachon, and others, are known even less well.

Candidate indicators for monitoring will include a subset most likely to provide the clearest "signal," alerting managers to the state of the system in time to respond with appropriate action. These "early-warning" indicators (NRC 1994) depend for their effectiveness on an understanding of the mechanistic behavior of the indicator in response to a specific stressor. Since the information necessary to guide and assure selection of the best indicators in all management scenarios is very seldom available, professional judgment must be used in their selection. Subsequent data collection will allow the assessment of the effectiveness of any given indicator. Similar to the development of conceptual models, input from participants in the adaptive-management process into the identification of programmatic measures allows different opinions regarding value and priority of managed resources to be considered in ecosystem planning.

These three essential activities that enable adaptive-management planning are essentially undeveloped on the Klamath River. Their absence holds back contributions from science needed to enhance the effectiveness of ecosystem management efforts. In contrast to the science generated in an adaptive-management program, science efforts in the Klamath River basin have often been reactionary, data collection and modeling being disconnected, and initiated in response to immediate management crises rather than developing coherent understanding or technical capabilities. Such a science agenda is hardly peculiar to the Klamath basin, but the approach is financially costly and ineffective in terms of providing management and policy insight and generating better understanding of the system. More important, science in reaction to crisis often does not focus on critical underlying uncertainties and may appear to be driven by the concerns of the party that sponsors the scientific effort. Reactive data collection and modeling tend to undercut stakeholder support of programmatic science, since scientific information seems to arrive too late to help resolve institutionalized management conflicts and sometimes does not provide clear choices when it does arrive (Figure 6-1).

Adaptive Management in Other Settings

Several approaches have been used to bring better knowledge to resource planning at regional spatial scales, similar to that of the Klamath River basin. Ecosystem-focused restoration programs that have been established for a decade or more offer organizational structures that explicitly integrate science into planning and project implementation and might serve as models for an integrated Klamath basin resource-management strategy. The Puget Sound Nearshore Partnership (www.pugetsoundnearshore.org), which shares salmon as a central management challenge with the Klamath basin, reviewed large-scale restoration programs in a search for models for integration of science into ecosystem management efforts; Figure 6-2 reproduces representative structures of four of them—the Chesapeake Bay Program; the CALFED Restoration program for the Sacramento and San Joaquin rivers, their tributaries, and the Sacramento River delta; the comprehensive Everglades Program; and the Glen Canyon Adaptive Management Program (see Van Cleve et al. 2004). All reviewed programs share characteristic vertical and horizontal integration of governance and responsibilities, with formal paths for moving information to the planning and policy process, and all follow various models of adaptive management. In each program, an adaptive-management working group (or implementation committee) made up of land and resource managers and technical experts in key management issue areas provides a central role, bringing scientific review and stakeholder input to inform both policy makers and management operations. The review identified characteristics of more-successful efforts; clearly articulated problem statements and program goals, independence of science activities from policy decisions, development of conceptual and quantitative models to resolve conflict and build scientific consensus, and identification of performance measures and initiation of monitoring efforts in an adaptive-management framework. Similar characteristics were identified by the NRC in salmon-restoration programs in the Pacific Northwest (NRC 1996).

Organizational attributes of these diverse ecosystem management approaches are actually shared with an effort ongoing in the Klamath River

204

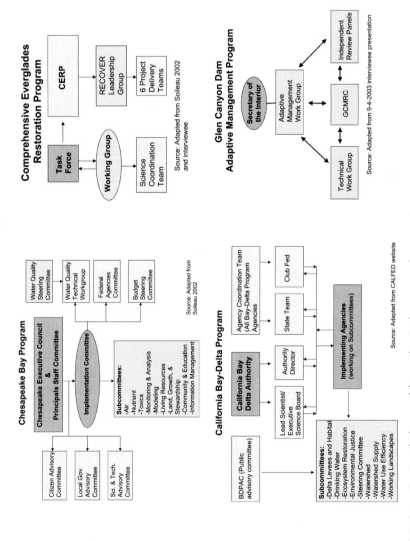

FIGURE 6-2 Organizational charts of four adaptively managed restoration programs.
SOURCE: Van Cleve et al. 2004. Reprinted with permission; copyright 2004, Puget Sound Nearshore Partnership.

basin. The Trinity River Restoration Program has harvested organizational features of successful programs elsewhere and is applying them in an effort to reverse severe degradation to the river's physical and biotic resources that has resulted from water diversions and dams.

Dissatisfied with a standing task force of government agency and stakeholder representatives, an implementation plan (established as the "preferred alternative" in the Final Trinity River Mainstem Fishery Restoration EIS/EIR [2000]) provides a governance structure that is explicitly intended to facilitate the program's Adaptive Environmental Assessment and Management efforts toward "learning by predicting the outcomes of management actions, implementing those actions, evaluating results, and rapidly improving future management decisions (Schleusner 2006). Included in the structure are a management council charged with policy and decision making; an adaptive-management working group, which facilitates stakeholder input to management recommendations; an adaptive-management assessment and management team of resource specialists and scientists, which designs and implements restoration projects and coordinates interagency activities and monitoring; and an independent scientific review board, which provides review of study plans, flow recommendations, restoration actions, and monitoring efforts. This governance structure appears to provide clear paths for bringing information that is critical to land, water, and species management to those who can use it. Adaptive management in the greater Klamath River basin would benefit substantially by adopting organizational and process approaches that are being used to support restoration planning in the Trinity River sub-basin and could enjoy enhanced effectiveness and efficiencies by collaborating with existing Trinity River efforts in a basin-wide science program.

The Klamath Basin's Conservation Improvement Program

In response to the perceived need for organized, coordinated information gathering and restoration efforts in the Klamath basin, the Bureau of Reclamation has attempted to organize and implement a Conservation Implementation Program (CIP) (USBR 2007b). According to the USBR, "the CIP is a mechanism by which participants will work together to

- Restore the Klamath River basin ecosystem.
- Further fulfill tribal trust responsibilities of the federal government.
- Allow continued, sustainable use of water.
- Foster lasting partnerships between governments and private stakeholders.

The CIP is intended "to coordinate conservation and restoration efforts throughout the Klamath River basin and provide technical and funding resources to achieve Klamath River basin ecosystem restoration and water management goals." The CIP also is intended to foster partnerships among government and private interests for the purpose of "restoring the Klamath River system" and sustaining "agricultural, municipal, and industrial water use, while reducing demand throughout the Klamath River basin." The CIP proposes to "fund research to increase understanding of the Klamath River system and monitoring to evaluate progress toward" program goals. In other words, the CIP intends to meet demonstrated needs in the Klamath basin for enhanced science.

Progress has been slow in the early stages of the endeavor; a program-description document is still in draft form after 3 years of discussion (C. Karas, USBR, personal communication, 2007). Unfortunately, because the effort to organize the CIP has been led by the USBR, the program, perhaps unfairly, is viewed by some stakeholders as not being independent of agency interests and biases. To be successful, the CIP must overcome long-standing issues of trust, authority, and prerogative. Some stakeholder groups appear to prefer a "bottom-up" organizational approach that empowers existing sub-basin working groups, less bureaucracy than the CIP currently involves. But having independent and insulated working groups has contributed (if not led) to the current lack of coordination among sub-basin programmatic actions and scientific efforts. Bottom-up organization is unlikely to produce integrated science, whole-system modeling, and efficiencies of scale in research and monitoring, as well as in providing the essential capacity for centralized information acquisition, synthesis, and dissemination. The resulting science will be unlikely to provide the answers to management questions that concern ecosystem management for the Klamath basin as a whole. Resolving this issue, that is, deciding whether to adopt a bottom-up or a top-down approach, or both, is critical. A working session on the CIP's organization has been postponed from early December 2006 to an as-yet unscheduled time (USBR 2007b).

Whatever the programmatic delivery system for scientific findings, a basin-wide science plan in support of adaptive management needs to be clearly articulated. The following design elements and goals are important:

1. Systematically reduce the uncertainties that limit the abilities of land and resource managers to operate the Klamath system to meet policy goals.

2. Show explicitly how knowledge derived from scientific efforts in the basin can be used to enhance the effectiveness, efficiency, and accountability of management actions and the policies on which they are based.

3. Develop and disseminate quantitative and conceptual models that describe the Klamath River basin and its attributes and functions to advance a general understanding of the ecosystem and the challenges to its management.

4. Identify performance measures that allow assessment of management actions, the state of the system, and trends of key resources.

5. Take advantage of flexibility in water management to test alternative specific flow options, which have been identified in conceptual models as having high likelihoods of contributing to salmon recovery.

6. Establish a periodic science forum at which the current state of knowledge of the Klamath basin system and its resources is promulgated and assessed, providing a basis for planning and prioritizing future scientific activities and approaches.

7. Produce periodically a volume or other written statement of the state of the Klamath River basin and the status and trends of its key resources of concern.

Science in the Klamath Basin

The models reviewed in this volume have had negligible positive effects from the viewpoints of key basin stakeholder groups. A USBR scientist repeated the statement that ecological "models have devastated farmers" in the upper Klamath basin; although, presumably, it has been management decisions based on the outputs of some models that are viewed unfavorably. But, accordingly, the pervasive distrust of federal authorities in the Klamath system has been joined with a distrust of science, and that distrust inevitably spills over, affecting stakeholder opinions of the reliability and independence of scientific efforts, such as the modeling efforts reviewed by this committee that are intended to address directly key management issues.

A structured science plan that is implemented adaptively should deliver products that are as value-neutral as possible, even while they address solutions to management problems. Those science products should not prescribe or recommend policy decisions, but should provide management authorities with knowledge that allows for better-informed decisions, and the course of implementing those decisions should be determined adaptively. Science that is problem-driven and that, by design, provides information that is directly useful in river operations and ecosystem restoration efforts should be emphasized.

Current science efforts in the Klamath basin show little of that applied focus or coherence. Although focused studies are numerous, and many of them are well designed and are producing useful information, there is no comprehensive research plan, little effort to develop larger coherent understanding from smaller scientific studies, and few clear links between science and application in management and restoration actions. Agency sponsors of Klamath basin research say they want to see it "less ad hoc and more systematic," and note they that "management coordination is getting better." Without a comprehensive management and research strategy, however, Klamath basin science will continue to have marginal utility and limited direct application, and management decisions will continue to suffer from a lack of informed guidance.

Science in support of ecosystem management in the Klamath River basin should have several features:

• Scientific efforts must be independent of political meddling. Although the science agenda needs to respond to the information needs of policy makers and land and resource managers, and information that may provide the platform for research and monitoring may come from diverse sources—including stakeholders with well-established positions regarding resources—the design of the science agenda and the experimental frameworks, data collection, analysis, and publications that result should be demonstrably the products of deliberations among scientists with appropriate quality controls and review. The scientists involved should be able to show that they have no explicit personal interest in management outcomes, and some reasonable balance in participating scientists from academia and agencies should be sought.

• Science needs to be institutionalized as a basin-wide program through which research, monitoring, and modeling efforts are organized, communicated, and subjected to independent review. A multidisciplinary science committee should be convened, with tasks that include producing a science plan that is periodically updated to be relevant and timely, and to reflect information needs of management agencies and advancements in the understanding of the Klamath River system and the status of key resources. This science committee should be more than a collection of scientists that represent particular interests in management of the basin.

• Leadership for the science program is necessary. A lead scientist with a history of objectivity and production of high-quality scientific products, as well as management and diplomatic skills, should represent scientific efforts in the Klamath River basin and serve as spokesperson for science to policy makers, management and resource agencies, key stakeholder groups, and the interested public.

• The products of scientific deliberations and activities, including research, monitoring, and modeling, must be as transparent as possible. The reasoning behind specific data gathering and analytic approaches needs to be clear, describable, and repeatable. Externally reviewed report and journal publication should be one desired end point of most basin science.

• Links between approaches and findings and their application in land, water, and resource management need to be clearly articulated and conveyed to managers and policy makers.

CONCLUSIONS AND RECOMMENDATIONS

Connecting science with public policy that produces sound decisions is a daunting challenge for the Klamath River basin, but the committee has identified two sets of conclusions and recommendations that represent a road map to success.

Conclusion 6-1. Planning for management and restoration of hydrologic and ecological research in the Klamath River basin is piecemeal, the Natural Flow Study being adopted for uses for which it was not originally intended and the Instream Flow Study being buffeted by critical comment from stakeholders. There is no overall independent coordination of science as it interacts with decision making.

Recommendation 6-1. A formal science plan for the Klamath River basin should support policy and decision making for the basin's hydrologic and ecological resources. Such a plan should prioritize data needs, identify key uncertainties, specify limits to management capabilities, conduct independent scientific review of research and management plans using that research, construct and oversee monitoring of the systems, and create hydrologic and ecological models.

Conclusion 6-2. There are no clearly defined connections between the conduct of science to understand the hydrology and ecology of the Klamath River basin and the conduct of decision making for resource management.

Recommendation 6-2. Planning for management and restoration of hydrologic and ecological resources in the Klamath River basin should use the formal science plan outlined in Recommendation 6-1 in an adaptive-management approach. This approach requires institutional structures and relationships that clearly designate specific authorities for policy implementation, as well as an adaptive-management working group that is an independent science-management team with representation of stakeholders, scientists, and resource management experts.

SUMMARY

More than 5 years after water deliveries were halted to farmers in the upper Klamath River basin in an attempt to get water to imperiled fishes downstream, and then, soon thereafter, salmon in the lower Klamath River experienced an unprecedented mass mortality, critical questions in Klamath basin ecosystem management remain unanswered. Is the federally listed coho salmon habitat-limited in the lower Klamath River and its tributaries? Do releases of water from the upper basin actually have beneficial effects on the lower basin fishery? These and other questions may only be addressed with coordinated modeling and experiments carried out in conjunction with ongoing management and restoration efforts, as authorities address the three central challenges for Klamath basin operations—assuring water deliveries, a sustainable salmon fishery, and persistence of listed species. Each is dynamic in nature and requires adaptive-management approaches that, in turn, require a programmatic approach that recognizes often distinct and sometimes complementary roles of decision makers, stakeholders, managers, and scientists.

Models of nature simply cannot capture all the breadth, complexity, and intricacy of a river system. No model of the hydrology and ecology of the Klamath River basin can possibly convert the currently limited available database into a finely resolved predictive tool. However, unguided intuition and political processes are even less likely to accomplish the objectives for water management in the Klamath basin. It is not unreasonable to conclude that many of the more critical shortcomings of the Natural Flow Study and the Instream Flow Study models reflect institutional impediments to the delivery of science to management that were faced by the modeling teams. From constraints placed on the modeling efforts at the outset, to interim applications of the models in unintended circumstances, to difficulties in accessing the best available information, circumstances conspired to compromise the content and utility of the final model products. Until the diverse and only weakly coordinated institutions that are charged to maintain and restore the Klamath River basin's ecosystem can find a means of effectively supporting scientific endeavors and applying the products of science, even the most reliable knowledge will remain ad hoc and relegated to the margins of river system planning and management. Management of the basin will then remain guided by less coherent and more controversial scientific activities and therefore more guided by political objectives with intuitive understandings of the system.

7

Conclusions and Recommendations

In the preceding chapters, the committee has explored in detail the Natural Flow Study (USBR 2005) and the Instream Flow Study (Hardy et al. 2006a) and assessed the scientific validity of the models produced by each study. Because the overall objective of the U.S. Bureau of Reclamation (USBR) has been to use the products of these studies to determine target flows for the Klamath River that can benefit coho and chinook salmon, the committee also evaluated the utility of the models and their outputs in potential support of regulation of river flows. In this chapter, the committee identifies the common threads that arose from its investigations and reviews the conclusions and recommendations that emerged from its assessments.

THE BIG PICTURE

The committee's considerations of science and decision making in the Klamath River basin identified the same overarching concern at almost every turn. The committee found that science was being carried out piecemeal, sometimes addressing very important questions, but not linking them to other relevant questions and studies. The Natural Flow Study (USBR 2005) and the Instream Flow Study (Hardy et al. 2006a) were major science and engineering investigations, but the linkage of one to the other was only partially achieved. Other studies in the basin, such as the U.S. Geological Survey's (USGS's) hydrologic studies in the Sprague River basin and the extensive research in the Trinity River basin (both of which are part of the

Klamath River basin), seem not to have had any influence on each other or on the flow studies examined in this report. The committee found that important characteristics of research for management of a complex river basin were missing from Klamath River studies: the need for a "big picture" perspective based on a conceptual model encompassing the entire basin and its many components. As a result, the integration of individual studies into a coherent whole has not taken place, and it is unlikely to take place under the present conditions. It also is not clear how much influence previous reports that have argued for integration (for example, Braunworth et al. 2002, IMST 2003, NRC 2004a) have had.

As shown in the previous chapters and summarized below, the committee found shortcomings in the Natural Flow and Instream Flow studies that are sufficiently serious that the committee questions whether the studies can guide decision making effectively. To address science and management in the basin, the committee first recommends that the agencies, researchers, decision makers, and stakeholders together define basin-wide science needs and priorities. One method of achieving success in this effort would be through the establishment of an independent entity to develop an integrated vision of science needs. The body that defines this vision must be viewed by all parties as truly independent for it to be effective, unlike the Conservation Improvement Program, which, despite good intentions, appears to many people in the region as a creature of the USBR and is therefore to be associated with the Bureau's official mandates and responsibilities. If the proposed task force reports to the secretary of the U.S. Department of the Interior (DOI), rather than to any specific agency, it is more likely to avoid the appearance of being controlled by any particular agency or interest group in the basin and thus is more likely to be and to appear independent. Leadership of the task force by a senior scientist who reports to the secretary would be a major step toward removing perceived biases in science and its application.

The committee concludes that when the science needs for the Klamath River basin are better characterized, the individual studies necessary to create a sound, science-based body of knowledge for decision makers and managers will be more easily identified. Only if this general vision and process determines that the Natural Flow Study and the Instream Flow Study might help to satisfy science needs in the basin should investigators seek to address the shortcomings that the committee has identified. The organizational structure and process by which the Trinity River Restoration Program was intended to implement science are sound, but in practice the implementation has been difficult due to a variety of challenges (Trinity Management Council Subcommittee 2004). Nonetheless, many aspects of the structure and process used on the Trinity River could be applied to the Klamath River.

FOUR MAJOR THEMES

Four major threads or themes arose from the committee's review of the two flow studies and their utility as scientific support for decision making: scale, the representativeness of data, connectivity, and a river-basin perspective. First, the issues of water resource use, ecosystem maintenance, and preservation of endangered species are multi-scalar, so any approach to the resolution of these issues also must be at multiple scales. For example, an understanding of the annual flow of water through the river system requires an appreciation of the effects of hemispheric climate systems on a watershed several thousand square kilometers in extent operating with changes over decades. On the other hand, an understanding of the geomorphic structure of the river channel that determines the habitat conditions experienced by fish requires appreciation of physical processes that operate at the scale of a few meters and change over periods of hours to a few days. Managers must decide water allocations and dam releases within the context of the laws and agreements that have histories of a century or more but must balance those decisions against political and social values that change over decades and in a context where drought or flood conditions present monthly dilemmas. The decisions about the water resources strongly affect local resource users, yet the fisheries resources are of much broader interest to stakeholders who are a national constituency. The committee found that the issue of scale of analysis pervaded its reviews of the two studies and their application and that clear specification of scale was critical to the potential success or failure of the studies.

The theme of scale gives rise to the second theme, representativeness. As an example, high-quality historical flow data for the Klamath River are available for only part of the twentieth century. These data were the foundation for many of the conclusions that went into the construction of the two flow models, but whether or not those data are representative of the entire twentieth century was not demonstrated. The Instream Flow Study could not examine habitats along the entire length of the Klamath River, so it was necessary to select a few relatively small reaches of the river for intensive evaluation. If these reaches are truly representative of the full length of river, the studies have much greater value than if the reaches are less representative. Representativeness is also a thread in the history of water management of the river. During most of the history of USBR management, agency objectives have been strongly oriented toward benefits for farmers and ranchers, without representation of the interests of Native Americans in the lower Klamath River basin or of the interests of protected or valued species. Now the agency has broadened its representation of these varied interests.

The third theme—connectivity (or the lack of it)—is an important characteristic of the Klamath River, and therefore it should be integral to the science used to understand the river and the decision-making processes used to manage the river. Many changes that have been imposed on the river system by human activities have decreased the connectivity of its various parts. For example, upstream from the Link River Dam and the body of Upper Klamath Lake extensive marshes, formerly functionally connected to the groundwater system and the surface flows, were drained and converted to agricultural use. In the river reach above the Keno Dam, construction of a railroad embankment disconnected the Klamath River from its historical flood overflow into Lower Klamath Lake; similarly, historical overflow into the Lost River also was controlled.

Connectivity also is important in the conduct of science. Science serves decision makers and the public most effectively if decision makers clearly define the purpose of the scientific investigations. If this purpose is unclear or if it changes during the course of research, the scientific products are likely to be less useful and may be wrongly or inappropriately applied. Connectivity—in the form of continuous communication—among the researchers themselves is crucial to the success of the scientific enterprise so that one part of the research (such as hydrology) provides inputs that are useful for another part (such as ecology). For this reason, frank and supportive communications between researchers are essential for outcomes of research to be useful to end-users. Connectivity among decision makers and stakeholders also is a fundamental prerequisite for the effective use of science in management. If decision makers and stakeholders lack common ground for exchange of ideas and resolution of conflicts, science is likely to be wasted, even if it is appropriate to the problem at hand, and management is likely to be fragmented. As a result, advocacy of limited perspectives is more likely to control outcomes, rather than compromises based on a sound understanding of the river's hydrologic and ecological processes.

Finally, successful science and decision making depend on a river-basin perspective. The human population of the Klamath basin is distributed over a landscape that includes parts of two states, several counties, and many communities. People living in the upper reaches of the basin have different livelihoods and expectations of Klamath River resources than the livelihoods and expectations of many residents of the lower reaches of the basin. The physical landscape and hydrologic and biological resources of the upper and lower basins are different from each other in fundamental ways. It is tempting to deal with the Klamath River basin for science and decision making from at least the standpoint of the upper and lower parts, but successful science and effective decisions are most likely to be the result of viewing the whole basin and all its parts together. This view includes taking into account distant parts of the basin when considering focused problems in limited areas. In the case of the Klamath River, for example, analysis of flows without considering upstream tributaries or tributary processes along the main stem neglects important explanations for river behavior, because that behavior is a product of all upstream processes.

NATURAL FLOW STUDY

The Natural Flow Study (USBR 2005) for the Klamath River has several admirable attributes. The river system is highly complex, and the flows of the river at the gauge near the Iron Gate Dam site reflect the influence of a complicated hydrologic system. The data sets describing stream flow that the Natural Flow Study assembled are extensive and are highly useful. The data adequately reflect the monthly seasonality of the flow system. Human activities have modified that system over substantial portions of the basin above the Iron Gate Dam gauge site, and USBR investigators included many of these modifications in their calculations. The investigators recognized the importance of marsh conversions and agricultural activities in affecting river flows and included these factors in their calculations. The documentation for the Natural Flow Study (USBR 2005) is accessible to the reader and provides a straightforward explanation of what the modelers did and how they did it and provides the complete output of the research. The report also addresses important issues about the natural flow model, including brief accountings of model verification, sensitivity, and uncertainty. As a result, it has some utility in providing a generalized picture of unimpaired (natural) flows in the system and in providing a general sense of minimum flows that should be provided to ensure the safety of the basin's fishes, although not precisely enough to lead to day-to-day management of the system.

The committee concluded, however, that the Natural Flow Study was compromised by the following fundamental issues, including the choice of a basic approach to natural flows, choices of the models for calculations, and omissions of factors likely to influence river flows at the Iron Gate Dam gauge site:

- The products of the Natural Flow Study, flow values for the Klamath River at the Iron Gate Dam site, were calculated as monthly values. However, ecological applications of the model require daily values (as discussed in more detail below in the section on the Instream Flow Study). As a result, the output of the Natural Flow Study would not have satisfied its ultimate use requirements even if the study had been executed without other errors.
- The basic approach used by USBR researchers to estimate the flows of the river without the upstream influence of dams and withdrawals relied on a "black box" method of accounting for flow using a standard spreadsheet as the foundation. While such an approach allowed ready calculations and simplicity of output, the approach is not supported by a

general understanding of physical processes that influence river flows. A physically based model that has seen widespread successful use elsewhere is the USGS's Modular Modeling System (MMS). The MMS provides greater flexibility and adaptability, and provides a firmer theoretical foundation than a straightforward accounting system.

• Calculations of the fate of water in the upper basin related to evapotranspiration were not done according to the best current methods. In constructing the mass budgets needed for estimations of natural flows, the Natural Flow Study correctly recognized the importance of evapotranspiration in the upper basin. Greater amounts of evapotranspiration in the upper basin result in lesser amounts of stream flow at the Iron Gate Dam site, and evapotranspiration is likely to change as a result of land use and land cover, particularly the installation of agricultural practices in place of natural vegetation. The USBR used the Soil Conservation Service (SCS)[1] modified Blaney-Criddle method for determining evapotranspiration from various land surfaces, but this method is now seriously outdated. A more recent and more sophisticated version of the method—the United Nations Food and Agriculture Organization's (FAO's) version of the modified Blaney-Criddle method—has improved accuracy for evapotranspiration calculations. The FAO modified Blaney-Criddle method has substantial data requirements, but all the required data are in the public domain and are easily accessed. Use of an up-to-date model would lend credibility to the natural flow estimates and would take advantage of already-collected data.

• The USBR (2005) Natural Flow Study attempted to calculate flows at Iron Gate Dam without addressing several important controlling factors for those flows. Groundwater plays a critical role in the hydrologic cycle of the upper Klamath River basin. Before the advent of agriculture, the exchange between groundwater and surface waters occurred along the courses of tributary rivers and in lake basins, sometimes through the intermediary zones of marshes and similar wetlands. After the introduction of agriculture, groundwater pumping and marsh drainage for fields and pastures became common, so the entire groundwater–surface-water connection was altered. Present groundwater-surface-water interactions therefore are highly unlikely to be similar to the connections that previously influenced natural flows. The Natural Flow Study did not adequately take into account the role of groundwater in the system.

• More generally, the Natural Flow Study did not address the issue of changes in land use and land cover. While the study did account for marsh conversions to agriculture, there are other important land-use changes that the study did not assess. For example, the study did not assess logging for lumber and forest clearing for agriculture, but these changes in the upper

---

[1]Now known as the Natural Resources Conservation Service.

Klamath River basin are potentially important in influencing downstream flows. Such land-use and land-cover changes also are important along the main stem of the lower Klamath River on tributary streams, because logging activities on the steep slopes of the region are likely to increase sediment inputs to the main stem. Remotely sensed data regarding land-use and land-cover change are available and can be analyzed using geographic information systems. Inclusion of land-use and land-cover analyses in the Natural Flow Study would have increased confidence in the resulting calculations, because, if such changes are important, they would reflect their influence in the model output. If the changes are unimportant, that outcome could be convincingly demonstrated.

• The Natural Flow Study failed to adequately model the connection between the Klamath River and Lower Klamath Lake. Under unregulated conditions, high flows in the Klamath River main-stem channel were able to overflow a shallow divide, and water coursed into the Lost River and to Lower Klamath Lake. During low-flow conditions in the Klamath River main-stem channel, flows in the main river were not deep enough to overflow the divide, and Lower Klamath Lake was essentially cut off from the main river channel. The availability of this "escape valve" probably was important in the pre-development river flow from Keno downstream. In the first decade of the twentieth century, the construction of a large levee to support a railroad effectively eliminated the original connection between the Klamath River and Lower Klamath Lake. Thereafter, high flows on the main stem of the Klamath River did not divert much water to the lake, leaving it in the main river. Even if all other things remained equal, the alteration of the Lower Klamath Lake connection would result in changed high flows at Keno and downstream. The hydrologic effects of this connection and its consequent elimination were poorly modeled with a regression function that mixed data from years before the disconnection and after it. Because those data were at monthly intervals, the model was further made unlikely to capture important dynamics of this hydrologic interaction. The inadequate and coarse-grained modeling of such a potentially important interaction reduces the utility of the natural flows calculated by USBR (2005).

• The Natural Flow Study did not adhere closely enough to standard scientific and engineering practice in the areas of calibration, testing, quality assurance, and quality control. These activities are prerequisites for confidence in the model products by users, including decision makers and other modelers.

The committee concluded that the Natural Flow Study includes calculated flows that are at best first approximations to useful estimates of such flows. The present version of the Natural Flow Study is less than adequate for input to the Instream Flow Study and for day-to-day decision making regarding flows to benefit the listed and other anadromous fish species in the Klamath River downstream from Iron Gate Dam. To become more useful for meaningful decision making in flow management, the Natural Flow Study should be improved by (1) replacing the SCS modified Blaney-Criddle method for calculating evapotranspiration with a more accurate and modern version, such as the FAO version of the method, using generally available data; (2) including groundwater dynamics in the model in at least a general way; (3) improving the portions of the predictive model relating to land use and land cover so that changes in these variables are represented in a more complete fashion; (4) explicitly modeling the connection between the Klamath River and Lower Klamath Lake and between the Klamath and Lost rivers during flooding and recession of floods; (5) replacing the black-box accounting method based on a spreadsheet with a more robust physically based model for generating flows, such as the USGS's MMS, or the new GSFLOW model, which couples with MMS and the groundwater model MODFLOW; (6) including an extensive investigation of high flows along with their geomorphic and ecological implications; and (7) adhering more closely to standard scientific and engineering practice by extensively validating and testing the models, while addressing issues of quality assurance and quality control. The set of natural-flow models used by decision makers must deal with the apparent paradox in the present model, whereby increased agricultural areas upstream produce increased river flows downstream. Useful models either will not produce such a result, or they will lend themselves to explanation of this counterintuitive result. Finally, and perhaps most fundamentally, if the NFS is to be used to support habitat and fish-population modeling as components of IFIM, output at a daily time step is needed.

INSTREAM FLOW STUDY

The Instream Flow Study (Hardy et al. 2006a) used products of the Natural Flow Study as inputs to a complex modeling project designed to connect river flows and channel characteristics with habitat suitability and fish populations. Several aspects of the study are praiseworthy. The measurement of stream-bed topography and substrate characteristics in this study represent innovative cutting-edge methods that provided generally useful representations of the river channel. The two-dimensional hydrodynamic model in the Instream Flow Study represented an improvement over one-dimensional flow models in simulating the hydraulic aspects of physical habitats. The application of two-dimensional approaches represented a willingness on the part of the investigators to engage in a highly complex and ambitious effort to deal with the hydraulic and hydrologic aspects of the problem of characterizing fish habitat. The study incorporated distance to escape cover, an important variable that is sometimes ignored in other studies. It leads to flow prescriptions that are closer in many aspects to natural flow patterns than the current flow regime. Model output includes comparisons of fish growth and the productivity of fish populations under current hydrologic conditions and for assumed hydrologic conditions with the flow recommendations in place. These analyses suggest some improvement in fish growth and production over current conditions, but they do not offer tools for evaluating tradeoffs between instream and out-of-stream uses of water.

As a general perspective, the Instream Flow Study followed the modular modeling process of Instream Flow Incremental Methodology (IFIM), which has seen wide application in studies of this type. Although in the committee's judgment, the IFIM approach to river-habitat studies often is improperly used, most particularly when the PHABSIM module output is used in isolation as a static index to generate a single, flat, minimum flow and therefore not satisfactory, the authors of the study addressed each of the component modules of the IFIM as a general process. They employed bioenergetics and a fish-population model to test their results, and they tried to compare observed with model-predicted fish locations.

Despite these strengths, the committee found important shortcomings in the Instream Flow Study and its use of models and data. Two shortcomings—use of monthly data and lack of tributary analyses—are so severe that that they should be addressed before decision makers use the outputs of the study. More fundamentally, the flow recommendations presented by the Instream Flow Study were not directly the result of physical-habitat modeling but rather reflect a sequence of estimations and comparisons among habitat values for various life stages derived from monthly flows and estimated monthly natural-flow values, interpolations, and the selection of the lower of either the natural monthly flow or a flow computed to provide the same amount of physical habitat as the natural flow. This series of adjustments led to flow recommendations that resembled the natural hydrograph in many aspects. These steps were not the result of systematic application of the IFIM method but instead resulted from multiple decisions. The recommendations were indirectly derived from the models and from the highly detailed three-dimensional imaging of habitats at the site scale that were collected for the models. Although the Instream Flow Study used the PHABSIM module and some aspects of the temperature and salmon-population modules of the IFIM, the approach taken has unique characteristics that need further testing to evaluate them. Improvements over the simple use of the PHABSIM output as an index of habitat quality were made by Hardy et al. (2006a).

The PHABSIM model has been criticized in the peer-reviewed literature, and improved sampling and statistical approaches to modeling habitat selection have been proposed (for example, Castleberry et al. 1996, Williams 1999, Guay et al. 2000, Kondolf et al. 2000, Manly et al. 2002, Railsback et al. 2003, Ahmani-Nedushan et al. 2006, Anderson et al. 2006). Members of this committee hold a variety of opinions on the degree to which PHABSIM incorporates current ecological and sampling theory and on the degree to which it can be relied on, even when it is applied with careful recognition of its constraints.

To the degree that any analysis (including that of Hardy et al. 2006a) relies on PHABSIM, it will need to convince others in the discipline that (1) all appropriate assumptions have been fully addressed; (2) the limitations of the model as documented in the scientific literature have been addressed; (3) both hydraulic and biological sub-models have been appropriately calibrated and tested against independent field data; and (4) the analysis recognizes that the hydraulic aspects of the habitat are but one element of a necessarily more comprehensive instream flow study. These matters are discussed in more detail below.

The authors of the Instream Flow Study (Hardy et al. 2006a) were provided only with monthly flow values by the USBR in the Natural Flow Study, although daily flows were recognized to be more useful. Monthly flow values can be useful for general river-basin planning, but they are not adequate for ecological modeling for river habitats, because the monthly average masks important discharge values that may exist only for a few days or even less. Sometimes these short-lived events may be over-bank flows, attended by important habitat expansions for fish, or they may be extreme low-flow events that can be detrimental to fish populations even if they last only a few days. These shorter-term variations in discharge can yield significant changes in stream hydraulics and temperature, both of which can have important ecological consequences. In either case, the very existence of critically important flow variations is masked by monthly averages, a fatal flaw. In short, planners may operate water systems on a monthly basis, but fish survive on a daily basis.

The elimination of consideration of tributary processes apparently resulted from an agreement reached by basin managers not to include tributary processes in the habitat studies, perhaps to simplify the engagement of stakeholders in the process. Since only the main stem of the Klamath River was subject to analysis, stakeholders with interests in tributary locations would not have to deal directly with the study. However, the river is a highly integrated hydrologic and ecological system. Its tributaries give the river some of its essential characteristics and provide some of the most important habitats in the basin. Detailed knowledge about the system—from the tributaries to the main channel downstream from Iron Gate Dam—is essential to the habitat analysis. The tributaries control the inflow of sediment and add important water to the main stem, they can provide important spawning and rearing habitats and serve as refuges for fish during some low-flow periods, and they influence water quality (sometimes positively, sometimes negatively). The Klamath River is not a confined gutter for rain-water, and therefore analyzing the river without considering its tributaries is akin to analyzing a tree by assessing only its trunk but not its branches. The previous NRC (2004a) report on the Klamath River basin also emphasized the importance of understanding the lower Klamath River tributaries and including them in restoration plans, especially for coho salmon.

The Instream Flow Study also exhibits modeling shortcomings. First, the study did not include important water-quality attributes, such as dissolved oxygen levels, nutrient loadings, contaminants, and sediment concentrations, each of which has important implications for the vitality of the fish populations of the Klamath River basin. Data on these attributes are sketchy in some cases, but at least a general assessment and discussion of the implications of water quality would have greatly enhanced the study.

Second, high flows are especially important to the physical and biological processes of the Klamath River, and further analysis of their frequency, duration, and timing is essential in understanding the dynamics of the river's hydrologic, geomorphologic, and ecological processes. High flows are agents of change in the morphology and substrate of the river, over-bank flows engage floodplains as elements of habitat available to fish, and high flows entrain and rearrange sediments within the channel. River-channel morphology is not static but rather adjusts to high flows, so channel change is not continuous but rather is an event-based process keyed to high flows. Reliance on monthly flow data, as noted above, made analysis of high flows impossible in the scope of the study.

Third, there was a lack of a thorough assessment of the relationship between flow-data time series and the behavior of different species and life stages and the population dynamics of coho and Chinook salmon. Such an analysis for both natural (historical) and existing flows would provide valuable insights into changes in the natural regime that have been brought about by human activities. It also would point the way toward evaluating alternative management scenarios capable of creating more natural river conditions that might lead to recovery of fish populations.

Fourth, the claim that the model outcomes are accurate, as assessed by some empirical tests of fish distributions and by use of bioenergetic and the SALMOD models, impairs the utility of the Instream Flow Study's prescriptions as representing the best alternative. Although the empirical tests and the bioenergetic and SALMOD model comparison to existing river-flow conditions suggest that the recommended flows offer some improvement over the current flow regime, they do not substitute for a rigorous statistical test of model predictions against observed distributions of fish and sensitivity analyses of changes in fish growth and population productivity related to changes in flow regime. Statistical measures of the closeness of fit between model predictions and fish occurrence would substantially increase confidence in the outputs of the study. When supported by sensitivity analyses, the demonstration of modeled fish growth and productivity as a consequence of alternative flow regimes would be a useful aid to decisions about water management.

Finally, there are three shortcomings in the experimental design of the Instream Flow Study: a fundamental beginning assumption about limits on salmon habitat, the representativeness of the reaches used for detailed study, and the statistical approach used to analyze the calculated set of instream flows. First, the study makes the implicit assumption that the primary limiting factor for the recovery of salmon is physical habitat, directly related to instream flows, but the study does not demonstrate when or even if physical habitat is a limiting factor in any of the life stages of the fishes of concern. The precise nature of any flow-related hydraulic-habitat "bottlenecks" in the population dynamics of the salmon is not demonstrated, so it is possible that temperature, dissolved oxygen, water quality, connectivity, disease, competition, or other factors are more critical to fish persistence than the hydraulic aspects of habitat are. In other words, suitable hydraulic-habitat conditions may be necessary but are not by themselves sufficient for fish persistence.

Second, the study used several relatively short reaches of the Klamath River for detailed analysis and testing of the model output because it was impossible to map and analyze in detail the entire length of the river from Iron Gate Dam to the sea. The selection of representative reaches seems reasonable, but the study does not justify the selection of the reaches used in the study and does not indicate how representative they are of the unstudied reaches. A cursory analysis of the entire river might be used to determine how representative the selected reaches truly are, but at present the representativeness of the selected reaches is unknown; therefore the utility of the results also is unknown.

Third, application of the Periodic Autoregressive Moving Average (PARMA) to analyze the calculated set of flows is problematic because the data were not normalized and spatial cross-correlations were not considered. Also, since a PARMA (5,0) model was used, the stochastic analysis does not reflect annual autocorrelation in the hydrologic data. To avoid compromising the reliability of model predictions, stochastic models must properly incorporate spatio-temporal correlation. By missing these attributes, the Instream Flow Study is seriously impaired.

The committee concludes that the study enhances understanding of the Klamath River basin ecosystem and the flows required to sustain it. In their present form, if they are adopted, the recommended flows resulting from the study should be adopted on an interim basis pending the model improvements outlined below to overcome its limitations, and a more integrated assessment of the scientific needs of the basin as a whole. The recommended flow regimes offer improvements over exiting monthly flows in that they include intra- and inter-annual variations and appear likely to enhance Chinook salmon growth and young-of-the-year production.

The committee recommends that the study be improved for greater utility by (1) using daily flows as a basis for calculations; (2) taking into account habitats, water, and sediment contributions from tributaries; (3) specifically testing how representative the selected test reaches are of the entire river; (4) rigorous statistical testing of the various model outcomes to support claims of accuracy; (5) including water quality measures, sediment loadings, and contaminants in the modeling process; (6) including extended analyses of high-flow events; (7) exploring through thorough analyses of the habitat time series the potential for improving conditions for a variety of species and life stages, assuming natural and existing flows and a series of possible alternative flows; (8) developing a more comprehensive stochastic model that reflects the spatio-temporal correlation of hydrologic processes acting in the basin; and (9) using dynamic fish-population growth and production models to investigate the influence of alternative flow regimes on life cycles and stages of salmon to determine the nature of potential habitat-related bottlenecks that can constrain population growth, as well as the potential for flow-related improvements.

WHAT IS THE UTILITY OF THE TWO
STUDIES FOR DECISION MAKING?

The committee has described the shortcomings of the Natural Flow and the Instream Flow studies as well as shortcomings imposed by the milieu in which scientific research in the Klamath River basin is planned, developed, and conducted. While these shortcomings limit potential applications of the two studies, they do not completely eliminate all potential model values. The Natural Flow Study, through its careful documentation and analysis, has provided a foundation that can be built on in future studies, and it has allowed some substantive insights to be developed. It has allowed a clearer vision of how the various parts of the Klamath River basin—especially areas above Upper Klamath Lake—interact with each other and with the Klamath River Project. Results from the Natural Flow Study facilitate the identification of additional information needs. The Instream Flow Study has made even clearer the importance of seasonal and inter-annual variability in stream flows to management and survival of the anadromous fishes of the river, and it has confirmed the apparent value of a seasonal flow pattern (hydrograph) in the Klamath River that resembles the shape of the natural hydrograph. It has helped to delimit the ranges of variability in stream flow that might be desirable and those that might not be tolerated by the fishes. It also has provided some insights into the ways that the anadromous fishes of the Klamath River use the various habitats it provides in different flow regimes.

Nonetheless, the two studies do not allow for a detailed and practical analysis of trade-offs among various flow-management regimes with respect to benefits and costs to the anadromous fishes in the river and to the agricultural and other interests in the basin. Before these system models can be used to guide management more specifically and with greater confidence, a more effective capacity for integrating the elements of the scientific endeavor in the basin will be needed, and the models' more important shortcomings will need to be addressed. The most critical shortcomings of the Natural Flow Study are its inadequate treatment of linkages between the Klamath River and Lower Klamath Lake, and its provision of only monthly, rather than daily, time steps for hydrologic data. For the Instream Flow Study, the most critical shortcomings are its lack of analysis of the Klamath River's tributaries and its use of monthly, instead of daily, flow values.

CONNECTING SCIENCE WITH DECISION MAKING

Connecting effective science with successful decision making for delivering water to users, sustaining downstream fisheries, and protecting the populations of protected species have been problematic in the Klamath River basin. The Natural Flow Study (USBR 2005) and the Instream Flow Study (Hardy et al. 2006a) are not likely to contribute effectively to sound decision making until political and scientific arrangements in the Klamath River basin that permit more cooperative and functional decision making can be developed. The employment of sound science will require the following elements:

1. A formal science plan for the Klamath River basin that defines research activities and the interconnections among them, along with how they relate to management and policy.

2. An independent science review and management mechanism that is isolated from direct political and economic influence and that includes a lead scientist or senior scientist position occupied by an authoritative voice for research.

3. A whole-basin viewpoint that includes both the upper and lower Klamath River basins with their tributary streams.

4. A data and analysis process that is transparent and that provides all parties with complete and equal access to information, perhaps through an independent science advisory group.

5. An adaptive management approach whereby decisions are played in water management with modeling efforts capable of evaluating alternative flow-management schemes and with monitoring and constant assessment, including assessment of any management actions taken and with occasional informed adjustments in management strategies.

The committee recommends that the researchers, decision makers, and stakeholders in the Klamath River basin evaluate the DOI-approved implementation plan for the Trinity River Implementation Program and emulate their counterparts in the Trinity River basin in attempting to connect scientific efforts and decision making and that the two units coordinate their research and management for the greater good of the entire river system.

References

Addley, R.C. 2006. A Mechanistic Daily Net Energy Intake Approach to Modeling Habitat and Growth of Drift-Feeding Salmonids (Particularly the Genus *Oncorhynchus*). Ph.D. Dissertation, Department of Civil and Environmental Engineering, Utah State University, Logan, UT.

Ahmadi-Nedushan, B., A. St-Hilaire, M. Bérubé, É. Robichaud, N. Thiémonge, and B. Bobée. 2006. A review of statistical methods for the evaluation of aquatic habitat suitability for instream flow assessment. River Res. Appl. 22(5):503-523.

Anderson, K.E., A.J. Paul, E. McCauley, L.J. Jackson, J.R. Post, and R.M.Nisbet. 2006. Instream flow needs in streams and rivers: The importance of understanding ecological dynamics. Front. Ecol. Environ. 4(6):309-318.

Anderson, M.P., and W.W. Woessner. 1992. Applied Groundwater Modeling: Simulation of Flow and Advective Transport. San Diego: Academic Press.

Annear, T., I. Chisholm, H. Beecher, A. Locke, P. Aarrestad, C. Coomer, C. Estes, J. Hunt, R. Jacobson, G. Jobsis, J. Kauffman, J. Marshall, K. Mayes, G. Smith, C. Stalnaker, and R. Wentworth. 2002. Instream Flows for Riverine Resource Stewardship. Cheyenne, WY: Instream Flow Council.

Annear, T., I. Chisholm, H. Beecher, A. Locke, P. Aarrestad, C. Coomer, C. Estes, J. Hunt, R. Jacobson, G. Jobsis, J. Kauffman, J. Marshall, K. Mayes, G. Smith, C. Stalnaker, and R. Wentworth. 2004. Instream Flows for Riverine Resource Stewardship, Rev. Ed. Cheyenne, WY: Instream Flow Council. 267 pp.

Antonio, D.B., and R.P. Hedrick. 1995. Effect of water temperature on infections with the microsporidian *Enterocytozoon salmonis* in Chinook salmon. Dis. Aquat. Org. 22(3):233-236.

Ayers Associates. 1999. Geomorphic and Sediment Evaluation of the Klamath River, California, Below Iron Gate Dam. Prepared for U.S. Fish and Wildlife Service, Yreka, CA, by Ayers Associates, Fort Collins, CO. March 1999.

Bartholow, J., J. Heasley, J. Laake, J. Sandelin, B.A.K. Coughlan, and A. Moos. 2002. SALMOD: A Population Model for Salmonids: User's Manual. Version W3. U.S. Geo-

226

logical Survey, Fort Collins, CO [online]. Available: http://www.fort.usgs.gov/products/publications/4046/4046.pdf [accessed June 26, 2007].

Bartholow, J.M. 1989. Stream Temperature Investigations: Field and Analytical Methods. Instream Flow Information Paper No. 22. Biological Report 89 (17). U.S. Fish and Wildlife Service, U.S. Department of the Interior, Washington, DC [online]. Available: http://www.krisweb.com/biblio/gen_usfws_bartholow_1989_br8917.pdf [accessed April 19, 2007].

Bartholow, J.M., and J.A. Henriksen. 2006. Assessment of Factors Limiting Klamath River Fall Chinook Salmon Production Potential Using Historical Flows and Temperatures. Open-File Report 2006-1249. Fort Collins Science Center, U.S. Geological Survey, Fort Collins, CO [online]. Available: http://www.fort.usgs.gov/products/publications/21759/21759.pdf [accessed June 26, 2007].

Bartholow, J.M., and T.J. Waddle. 1986. Introduction to Stream Network Habitat Analysis. Instream Flow Information Paper No. 22. Biological Report 86(8). Fish and Wildlife Service, U.S. Department of the Interior, Washington, DC. 242 pp.

Bartholow, J.M., J.L. Laake, C.B. Stalnaker, and S.C. Williamsom. 1993. A salmonid population model with emphasis on habitat limitations. Rivers 4(4):265-279.

Bartholow, J.M., J. Heasley, R.B. Hanna, J. Sandelin, M. Flug, S. Campbell, J. Hendriksen, and A. Douglas. 2003. Evaluating Water Management Strategies with the Systems Impact Assessment Model: SIAM, Version 3. Open File Report 03-82. Fort Collins Science Center, U.S. Geological Survey, Fort Collins, CO. March 2003 [online]. Available: http://www.fort.usgs.gov/products/publications/10015/10015.pdf [accessed April 20, 2007].

Bartolini, P., and J.D. Salas. 1993. Modeling of streamflow processes at different time scales. Water Resour. Res. 29(8):2573-2588.

Beaumont, M.W., P.J. Butler, and E.W. Taylor. 1995. Exposure of brown trout, *Salmo trutta*, to sub-lethal copper concentrations in soft acidic water and its effect upon sustained swimming performance. Aquat. Toxicol. 33(1):45-63.

Beck, M.B. 2002. Model evaluation and performance. Pp. 1275-1279 in Encyclopedia of Environmetrics, Vol. 3, A.H. El-Shaarawi, and W.W. Piegorsch, eds. Chichester: Wiley [online]. Available: http://www.cgrer.uiowa.edu/cleaner/italy/Beck_Venicem.pdf [accessed Feb. 12, 2007].

Beckham, S.D., and T.W. Canaday. 2006. Historical Landscape Overview of the Upper Klamath River Canyon of Oregon and California. Cultural Resource Series No. 13. Bureau of Land Management, U.S. Department of the Interior, Portland, OR.

Bidlake, W.R., and K.L. Payne. 1998. Evapotranspiration from Selected Wetlands at Klamath Forest and Upper Klamath National Wildlife Refuges, Oregon and California. Report prepared by U.S. Geological Survey for U.S. Fish and Wildlife Services, Tacoma, WA.

Bilby, R.E., B.R. Fransen, P.A. Bisson, and J.K. Walter. 1998. Response of juvenile coho salmon (*Oncorhynchus kisutch*) and steelhead (*Oncorhynchus mykiss*) to the addition of salmon carcasses to two streams in southwestern Washington, U.S.A. Can. J. Fish. Aquat. Sci. 55(8):1909-1918.

Bissonette, J.A. 1997. Scale-sensitive ecological properties: Historical context, current meaning. Pp. 3-31 in Wildlife and Landscape Ecology: Effects of Pattern and Scale, J.A. Bissonette, ed. New York: Springer.

Bovee, K.D. 1982. A Guide to Stream Habitat Analysis Using the Instream Flow Incremental Methodology. FWS/OBS-82/26. Instream Flow Information Paper No. 12. U.S. Fish and Wildlife Service, U.S. Department of the Interior, Washington, DC.

Bovee, K.D., B.L. Lamb, J.M. Bartholow, C.B. Stalnaker, J. Taylor, and J. Henriksen. 1998. Stream Habitat Analysis Using the Instream Flow Incremental Methodology. U.S. Geological Survey, Biological Resources Division, Fort Collins, CO [online]. Available: http://www.fort.usgs.gov/products/publications/3910/3910.pdf [accessed Feb. 22, 2007].

Boyle, J.C. 1976. 50 Years on the Klamath. Medford, OR: Klocker Printery. [Klamath County Museum, Klamath Falls, OR].

Braunworth, W.S., Jr., T. Welch, and R. Hathaway, eds. 2002. Water Allocation in the Klamath Reclamation Project, 2001: An Assessment of Natural Resource, Economic, Social, and Institutional Issues with a Focus on the Upper Klamath Basin. Special Report 1037. Corvallis: Oregon State University Extension Service [online]. Available: http://extension.oregonstate.edu/catalog/html/sr/sr1037/report.pdf [accessed Oct. 23, 2007].

Brett, J.R. 1964. The respiratory metabolism and swimming performance of young sockeye salmon. J. Fish Res. Board Can. 21(5):1183-1226.

Bridge, J.S. 2003. Rivers and Floodplains: Forms, Processes, and Sedimentary Record. Malden, MA: Blackwell.

Burgman, M.A., D.R. Breininger, B.W. Duncan, and S. Ferson. 2001. Setting reliability bounds on habitat suitability indices. Ecol. Appl. 11(1):70-78.

Burns, J.W. 1970. Spawning bed sedimentation studies in Northern California streams. Calif. Fish Game 56(4):253-270.

CALFED. 2000. Ecosystem Restoration Program Plan, Vol. 3. Strategic Plan for the Ecosystem Restoration. Final Programmatic EIS/EIR Technical Appendix. CALFED Bay Delta Program, Sacramento, CA. July 2000 [online]. Available: http://www.delta.dfg.ca.gov/erp/docs/reports_docs/ERPP_Vol_3.pdf [accessed Feb.12, 2007].

Castleberry, D.T., J.J. Czech, D.C. Erman, D. Hankin, M. Healey, G.M. Kondolf, M. Mangel, M. Mohr, P.B. Moyle, J. Nielsen, T.P. Speed, and J.G. Williams. 1996. Uncertainty and instream flow standards. Fisheries 21(8):20-21.

CDFG (California Department of Fish and Game). 1994. Petition to the California Board of Forestry to List Coho Salmon (*Oncorhynchus kisutch*) as a Sensitive Species. California Department of Fish and Game [online]. Available: http://www.dfg.ca.gov/nafwb/pubs/1994/coho_pet.pdf [accessed April 10, 2007] (as cited in Weitkamp et al. 1995).

CDFG (California Department of Fish and Game). 1995. Klamath River Basin Fall Chinook Salmon Run-Size, In-River Harvest Escapement—1995 Season. Klamath Trinity Program.

CDFG (California Department of Fish and Game). 2004. September 2002 Klamath River Fish-Kill: Final Analysis of Contributing Factors and Impacts. California Department of Fish and Game, California Resources Agency, Sacramento. July 2004 [online]. Available: http://www.pcffa.org/KlamFishKillFactorsDFGReport.pdf [accessed June 2006].

Cherry, D.S., S.R. Larrick, J.D. Giattina, K.L. Dickson, and J. Cairns, Jr. 1978. The avoidance response of the common shiner to total and combined residual chlorine in thermally influenced discharges. Pp. 826-837 in Energy and Environmental Stress in Aquatic Systems, J.H. Thorp, and J.W. Gibbons, eds. National Technical Information Service, U.S. Department of Commerce, Washington, DC.

Chow, V.T., D.R. Maidment, and L.W. Mays. 1988. Applied Hydrology. New York, NY: McGraw Hill.

City-Data.com. 2007. Weaverville-California [online]. Available: www.city-data.com/city/Weaverville-California.html [accessed April 10, 2007].

City of Klamath Falls. 2007. City of Klamath Falls, Oregon: City Profile [online]. Available: http://www.ci.klamath-falls.or.us/ [accessed April 10, 2007]

Coots, M. 1962. Shasta River, Siskiyou County, 1958 King Salmon Count with Yearly Totals From 1930-1961. Marine Resources Administrative Report No. 62-64. California Department of Fish and Game, Inland Fisheries Branch. 5 pp.

Coots, M. 1973. A Study of Juvenile Steelhead, *Salmo gairdenerii gairdernii* Richardson, in San Gregorio Creek and Lagoon, San Mateo County, March through August, 1971. California Anadromous Fisheries Branch Administrative Report No. 73-4. Sacramento: California Department of Fish and Game.

Crawford, N.H., and R.K. Linsley. 1962. The Synthesis of Continuous Streamflow Hydrographs on a Digital Computer. Department of Civil Engineering Technical Report 12. Palo Alto, CA: Stanford University.

Crawford, N.H., and R.K. Linsley. 1966. Digital Simulation in Hydrology: Stanford Watershed Model IV. Department of Civil Engineering Technical Report 39. Palo Alto, CA: Stanford University.

Cuenca, R.H. 1989. Irrigation System Design: An Engineering Approach. Englewood Cliffs, NJ: Prentice Hall.

Cuenca, R.H., J.L. Nuss, A. Martinez-Cob, G.G. Katul, and J. McFaci-Ganzales. 1992. Oregon Crop Water Use and Irrigation Requirements. Corvallis, OR: Oregon State University.

Deas, M.L., S.K. Tanaka, and J.C. Vaughn. 2006. Klamath River Thermal Refugia Study: Flow and Temperature Characterization—Final Report. Watercourse Engineering, Inc., Davis, CA. March 6, 2006. 244 pp.

DHI. 2006. MIKE-SHE, Integrated Surface Water and Groundwater Model. DHI Software [online]. Available: http://www.dhigroup.com/Software/WaterResources/MIKESHE.aspx [accessed Feb. 12, 2007].

Dietrich, W.E. 1987. Mechanics of flow and sediment transport in river bends. Pp. 179-227 in River Channels: Environment and Process, K. Richards, ed. New York: Basil Blackwell.

Dietrich, W.E., and J.D. Smith. 1983. Influence of the point bar on flow through curved channels. Water Resour. Res. 19(5):1173-1192.

Dingman, S.L. 1989. Probability distribution of velocity in natural-channel cross sections. Water Resour. Res. 25(3):508-518.

Donigian, A.S., Jr., B.R. Bicknell, and J.C. Imhoff. 1995. Hydrologic simulation program – FORTRAN (HSPF). Pp. 395-442 in Computer Models of Watershed Hydrology, V.P. Singh, ed. Highlands Ranch, CO: Water Resources Publications.

Downer, C.W., F.L. Ogden, J. M. Niedzialek, and S. Liu. 2006. Gridded Surface/Subsurface Hydrologic Analysis (GSSHA) model: A model for simulating diverse streamflow producing processes. Pp. 131-159 in Watershed Models, V.P. Singh, and D. Frevert, eds. Boca Raton: CRC Press.

Drechsler, M. 1998. Sensitivity analysis of complex models. Biol. Conserv. 86(3):401-412.

Dunsmoor, L.K., and C.W. Huntington. 2006. Suitability of Environmental Conditions within Upper Klamath Lake and the Migratory Corridor Downstream for Use by Anadromous Salmonids. Technical Memorandum to the Klamath Tribes. October 2006.

Elliott, J.M. 1994. Quantitative Ecology and the Brown Trout. Oxford: Oxford University Press.

Enfield, D.B., A.M. Mestas-Nuez, and P.J. Trimble. 2001. The Atlantic multidecadal oscillation and its relation to rainfall and river flows in the continental U.S. Geophys. Res. Lett. 28(10):2077-2080.

EPA (U.S. Environmental Protection Agency). 1990. The Quality of Our Nation's Water: A Summary of the 1988 National Water Quality Inventory. EPA 440/4-90-005. Office of Water, U.S. Environmental Protection Agency, Washington, DC. May 1990.

EPA (U.S. Environmental Protection Agency). 2003. Draft Guidance on the Development, Evaluation, and Application of Regulatory Environmental Models. The Council for Regulatory Environmental Modeling, Office of Science Policy, Office of Research and Development, U.S. Environmental Protection Agency, Washington, DC [online]. Available: http://www.epa.gov/ord/crem/library/CREM%20Guidance%20Draft%2012_03.pdf [accessed Feb. 20, 2007].

EPA (U.S. Environmental Protection Agency). 2006. Watershed Priorities: Klamath River Basin, California & Oregon. U.S. Environmental Protection Agency, Region 9, Water

Program [online]. Available: www.epa.gov/region9/water/watershed/klamath.html [accessed Dec. 27, 2006].

EPRI (Electric Power Research Institute). 2000. Instream Flow Assessment Methods: Guidance for Evaluating Instream Flow Needs in Hydropower Licensing. Technical Report TR-1000554. Electric Power Research Institute, Palo Alto, CA. December.

Fennema, R.J., C.J. Neidrauer, R.A. Johnson, T.K. MacVicar, and W.A. Perkins. 1994. A computer model to simulate natural Everglades hydrology. Pp. 249-289 in Everglades: The Ecosystem and Its Restoration, S.M. Davis, and J.C. Ogden, eds. Delray Beach, FL: Lucie Press.

Fetter, C.W. 2001. Applied Hydrogeology, 4th Ed. Upper Saddle River, NJ: Prentice Hall. 598 pp.

Fleming, G. 1974. Deterministic Simulation in Hydrology. New York: American Elsevier.

Fletcher, T., and G. Addington. 2007.Klamath Basin Talks Best Hope for Future. Portland Oregonian August 13, 2007 [online]. Available: http://www.oregonlive.com/commentary/oregonian/index.ssf?/base/editorial/1186790138273190.xml&coll=7 [accessed October 2007].

Fortune, J.D., A.R. Gerlach, and C.J. Hanel. 1966. A Study to Determine the Feasibility of Establishing Salmon and Steelhead in the Upper Klamath Basin. Oregon State Game Commission and Pacific Power and Light.

Freeman, M.C., Z.H. Bowen and J.H. Crance. 1997. Transferability of habitat suitability criteria for fishes in warmwater streams. N. Am. J. Fish. Manage. 17(1):20-31.

Frey, H.C., and S.R. Patil. 2002. Identification and review of sensitivity analysis methods. Risk Anal. 22(3):553-578.

Gan, K., and T. McMahon. 1990. Variability of results from the use of PHABSIM in estimating habitat area. Regul. River 5(3):233-239.

Gannett, M. 2007. Ground-Water Hydrology of the Upper Klamath Basin. Presentation at the Third Meeting on Hydrology, Ecology, and Fishes of the Klamath River Basin, January 29-31, 2007, Klamath Falls, OR.

Gannett, M.W., K.E. Lite, Jr., and J.L. La Marche. 2003. Temporal and spatial variations in groundwater discharge to streams in the Cascade Range in Oregon and implications for water management in the Klamath River Basin. Geological Society of America Abstracts with Programs 35(6):487.

Garshelis, D.L. 2000. Delusions in habitat evaluation: Measuring use, selection, and importance. Pp. 111-164 in Research Techniques in Animal Ecology, Controversies and Consequences, L. Boitani, and T.K. Fuller, eds. New York: Columbia University Press.

Gass, S.I. 1983. Decision-aiding models: Validation, assessment, and related issues for policy analysis. Oper. Res. 31(4):603-631.

Geoffrion, A.M. 1976. The purpose of mathematical programming is insight, not numbers. Interfaces 7(1):81-92.

Ghanem, A., P. Steffler, F. Hicks, and C. Katopodis. 1996. Two-dimensional hydraulic simulation of physical conditions in flowing streams. Regul. River 12(2-3):185-200.

Graham, D.N., and M.B. Butts. 2006. Flexible integrated watershed modeling with MIKE SHE. Pp. 245-272 in Watershed Models, V.P. Singh, and D. Frevert, eds. New York: Taylor and Francis Group, CRC Press.

Grimm, V., and S.F. Railsback. 2005. Individual-Based Modeling and Ecology. Princeton: Princeton University Press.

Gross, M.R. 1987. Evolution of diadromy in fishes. Pp. 14-25 in Common Strategies of Anadromous and Catadromous Fishes, M.J. Dadswell, ed. American Fisheries Society Symposium No. 1. Bethesda, MD: American Fisheries Society.

Guay, J.C., D. Boisclair, D. Rioux, M. Leclerc, M. Lapointe, and P. Legendre. 2000. Development and validation of numerical habitat models for juveniles of Atlantic salmon (*Salmo salar*). Can. J. Fish. Aquat. Sci. 57(10):2065-2075.

Hamilton, J.B., G.L. Curtis, S.M. Snedaker, and D.K. White. 2005. Distribution of anadromous fishes in the Upper Klamath River watershed prior to hydropower dams—A synthesis of the historical evidence. Fisheries 30(4):10-20.

Hamilton, S.J., and K.J. Buhl. 1990. Safety assessment of selected inorganic elements to fry of chinook salmon (*Oncorhynchus tshawytscha*). Ecotoxicol. Environ. Saf. 20(3): 307-324.

Hanna, R.B., and S.G. Campbell. 2000. Water Quality Modeling in the Systems Impact Assessment Model for the Klamath Basin—Keno, Oregon to Seiad Valley, California. Open File Report 99-113. Fort Collins, CO: U.S. Geological Survey.

Hanson, P.C., T.B. Johnson, D.E. Schindler, and J.F. Kitchell. 1997. Fish Bioenergetics 3.0 Software for Windows. WISCU-T-97-001. University of Wisconsin Sea Grant Institute, Madison, WI.

Hardin, T.S., R.T. Grost, M.B. Ward, and G.E. Smith. 2005. Habitat Suitability Criteria for Anadromous Salmonids in the Klamath River, Iron Gate Dam to Scott River, California. Stream Evaluation Report 05-1. U.S. Fish and Wildlife Service and California Department of Fish and Game. March 2005. 73 pp [online]. Available: http://www.instreamflowcouncil.org/images/lr_stream_report_05-1.pdf [accessed March 14, 2007].

Hardy, T.B. 1999. Evaluation of Interim Instream Flow Needs in the Klamath River: Phase I. Final Report. Prepared for U.S. Department of the Interior, by Institute of Natural Systems Engineering, Utah Water Research Laboratory, Utah State University, Logan, UT. August 5, 1999 [online]. Available: http://www.usbr.gov/mp/kbao/esa/10_Hardy PhaseIReport_4_6_01.pdf [accessed April 19, 2007].

Hardy, T.B., and R.C. Addley. 2001. Evaluation of Interim Instream Flow Needs in the Klamath River: Phase II, Final Report. Prepared for U.S. Department of the Interior, by Institute for Natural Systems Engineering, Utah State University, Logan, UT. November 21, 2001 [online]. Available: http://www.krisweb.com/biblio/klamath_usdoi_hardy_2003_phase2draft.pdf [accessed April 11, 2007]

Hardy, T.B., R.C. Addley, and E. Saraeva. 2006a. Evaluation of Instream Flow Needs in the Lower Klamath River: Phase II, Final Report. Institute for Natural Systems Engineering, Utah State University, Logan, UT. July 31, 2006 [online]. Available: http://www.neng.usu.edu/uwrl/inse/klamath/FinalReport/PhaseII_Final_Report_Revised_Oct_16_2006.pdf [accessed April 11, 2007].

Hardy, T.B., T. Shaw, R.C. Addley, G.E. Smith, M. Rode, and M. Belchik. 2006b. Validation of Chinook fry behavior-based escape cover modeling in the lower Klamath River. Intl. J. River Basin Mgmt 4 (2):1-10.

Hay, L.E., G.H. Leavesley, and M.P. Clark. 2005. Use of remotely-sensed snow covered area in watershed model calibration for the Sprague River in the Upper Klamath Basin. Eos Trans. AGU 86(52), Fall Meet. Suppl., Abstract H31H-07.

Healey, M.C., P.L. Angermeier, K.W. Cummins, T. Dunne, W. J. Kimmerer, G.M. Kondolf, P.B. Moyle, D. D. Murphy, D.T. Patten, D.J. Reed, R.B. Spies, and R.H. Twiss. 2007. Conceptual Models and Adaptive Management in Ecological Restoration: The CALFED Bay-Delta Environmental Restoration Program. Environmental Management [online]. Available: http://calwater.ca.gov/Programs/EcosystemRestoration/ERP_Science_Board/Big_Model_Paper_5-3-04.pdf [accessed April 30, 2007].

Hecht, B., and G.R. Kamman. 1996. Initial Assessment of Pre- and Post-Klamath Project Hydrology on the Klamath River and Impacts of the Project on Instream Flows and Fishery Habitat. Prepared on Behalf of The Yurok Tribe, Eureka, CA. Balance Hydrologics, Inc., Berkeley, CA. March 1996 [online]. Available: http://www.balancehydro.com/pdf/LowerKlamathReport.pdf [accessed Feb. 20, 2007].

Heggenes, J. 1996. Habitat selection by brown trout (*Salmo trutta*) and young Atlantic salmon (*S. salar*) in streams: Static and dynamic hydraulic modeling. Regul. River 12(2-3):155-169.

Helfrich, L.A., C. Liston, S. Hiebert, M. Albers, and K. Frazer. 1999. Influence of low-head diversion dams on fish passage, community composition, and abundance in the Yellowstone River, Montana. Rivers 7(1):21-32.

Hickey, C.W., and L.A. Golding. 2002. Response of macroinvertebrates to copper and zinc in a stream mesocosm. Environ. Toxicol. Chem. 21(9):1854–1863.

Holling, C.S., ed. 1978. Adaptive Environmental Assessment and Management. New York: Wiley.

Hubbard, L.L. 1970. Water Budget of Upper Klamath Lake, Southwestern Oregon. Hydrologic Investigations Atlas HA-351. Washington, DC: U.S. Department of the Interior, U.S. Geological Survey.

HydroGeoLogic Inc. 1997. MODFLOW-SURFACT, Version 2.1, Code Documentation and User's Guide. HydroGeoLogic Inc., Herndon, VA. 231 pp.

IMST (Independent Multidisciplinary Science Team). 2003. IMST Review of the USFWS and NMFS 2001 Biological Opinions on Management of the Klamath Reclamation Project and Related Reports. Technical Report 2003-1 to the Oregon Plan for Salmon and Watersheds, Oregon Watershed Enhancement Board, Salem [online]. Available: http://www.fsl.orst.edu/imst/reports/2003-01.pdf [accessed June 2006].

Jackson, S. 2006. Water models and water politics: Design, deliberation, and virtual accountability. Pp. 95-104 in Proceedings of the 7th Annual International Conference on Digital Government Research, May 21-24, 2006, San Diego, CA. ACM International Conference Proceeding Series Vol. 151. New York: ACM Press.

Jensen, M.E., R.D. Burman, R.G. Allen, eds. 1990. Evapotranspiration and Irrigation Water Requirements. ASCE Manuals and Reports on Engineering Practice No. 70. New York: American Society of Civil Engineers.

Jobling, M. 1997. Temperature and growth: Modulation of growth rate via temperature change. Pp. 225-254 in Global Warming: Implications for Freshwater and Marine Fish, C.M. Wood, and D.G. McDonald, eds. Cambridge, UK: Cambridge University Press.

Jobling, S., M. Nolan, C.R. Tyler, G. Brighty, and J.P. Sumpter. 1998. Widespread sexual disruption in wild fish. Environ. Sci. Technol. 32(17):2498–2506.

Johnson, W.C. 1992. Dams and riparian forests: Case study from the Upper Missouri River. Rivers 3(4):229-242.

Kauffman, B., T. Bettge, L. Buja, T. Craig, C. DeLuca, B. Eaton, M. Hecht, E. Kluzek, J. Rosinski, and M. Vertenstein. 2001. Community Climate System Model Software Developer's Guide. Community Climate System Model Working Group. June 2001 [online]. Available: http://www.ccsm.ucar.edu/working_groups/Software/dev_guide/dev_guide/dev_guide.html [accessed Feb. 15, 2007].

Kennedy, A.M., D.C. Garen, and R.W. Koch. 2006. Improved early season seasonal streamflow volume prediction models considering equatorial sea surface temperature gradients: Trans-Nino Index. Geophysical Research Abstracts 8:02481. European Geophysical Union.

Kimmerer, W.J., D.D. Murphy, and P.L. Angermeier. 2005. A landscape-level model for ecosystem restoration in the San Francisco estuary and its watershed. San Francisco Estuary and Watershed Science 3(1):Article 2. March 2005 [online]. Available: http://repositories.cdlib.org/jmie/sfews/vol3/iss1/art2 [accessed Feb. 15, 2007].

Kleijnen J.P.C. 1995. Verification and validation of simulation models. Eur. J. Oper. Res. 82(1):145-162.

Knighton, D. 1998. Fluvial Forms and Processes: A New Perspective. New York: Arnold.

Kondolf, G.M. 2000. Assessing salmonid spawning gravel quality. T. Am. Fish. Soc. 129(1): 262-281.

Kondolf, G.M., and W.V.G. Matthews. 1993. Management of Coarse Sediment in Regulated Rivers of California. Report No. 80. University of California Water Resources Center, Riverside, CA.

Kondolf, G.M., E.W. Larsen, and J.G. Williams. 2000. Measuring and modeling the hydraulic environment for assessing instream flows. N. Am. J. Fish. Manage. 20(4):1016-1028.

Kondolf, G.M., A. Boulton, S. O'Daniel, G. Poole, F. Rahel, E. Stanley, E. Wohl, A. Bang, J. Carlstrom, C. Cristoni, H. Huber, S. Koljonen, P. Louhi, and K. Nakamura. 2006. Process-based ecological river restoration: Visualizing three-dimensional connectivity and dynamic vectors to recover lost linkages. Ecology and Society 11(2):5 [online]. Available: http://www.ecologyandsociety.org/vol11/iss2/art5/ [accessed Oct. 2007].

Konikow, L.F., and J.D. Bredehoeft. 1992. Ground-water models cannot be validated. Adv. Water Resour. 15(1):75-83.

Kumar, D.N., U. Lal, and M.R. Petersen. 2000. Multisite disaggregation of monthly to daily streamflow. Water Resour. Res. 36(7):1823-1834.

Labbe, T. R., and K. D. Fausch. 2000. Dynamics of intermittent stream habitat regulate persistence of a threatened fish at multiple scales. Ecol. Appl. 10(6):1774-1791.

Lamouroux, N., Y. Souchon, and E. Herouin. 1995. Predicting velocity frequency distributions in stream reaches. Water Resour. Res. 31(9):2367-2376.

Lamouroux, N., H. Capra, and M. Pouilly. 1998. Predicting habitat suitability for lotic fish: Linking statistical hydraulic models with multivariate habitat use models. Regul. River.14(1):1-12.

Larsen, E.W. 1995. The Mechanics and Modeling of River Meander Migration. Ph. D. Thesis, University of California, Berkeley.

Leavesley, G.H., S.L. Markstrom, and R.J.Viger. 2006. USGS Modular Modeling System (MMS)-Precipitation-Runoff Modeling System (PRMS). Pp. 159-178 in Watershed Models, V.P. Singh, and D. Frevert, eds. New York: Taylor and Francis Group, CRC Press.

Leclerc, M., A. Boudreault, T.A. Bechara, and G. Corfa. 1995. Two-dimensional hydrodynamic modeling: A neglected tool in the instream flow incremental methodology. T. Am. Fish. Soc. 124(5):645-662.

Lee, C.G., A.P. Farrell, A. Lotto, M.J. MacNutt, S.G. Hinch, and M.C. Healey. 2003. The effect of temperature on swimming performance and oxygen consumption in adult sockeye (Oncorhynchus nerka) and coho (O. kisutch) salmon stocks. J. Exp. Biol. 206(18):3239–3251.

Leidy, R.A., and G.R. Leidy. 1984. Life Stage Periodicities of Anadromous Salmonids in the Klamath River Basin, Northwestern California. U.S. Fish and Wildlife Service, Sacramento, CA. April 1984 [online]. Available: http://www.krisweb.com/biblio/klamath_usfws_leidyetal_1984.pdf [accessed April 10, 2007].

Leonard, P.M., and D.J. Orth. 1988. Use of habitat guilds of fishes to determine instream flow requirements. N. Am. J. Fish. Manage. 8(4):399-409.

Leopold, L.B., M.G. Wolman, and J.P. Miller. 1964. Fluvial Processes in Geomorphology. San Francisco: W.H. Freeman.

Levin, S.A. 1992. The problem of pattern and scale in ecology. Ecology 73(6):1943-1967.

Lisle, T.E., J.M. Nelson, J. Pitlick, M.A. Madej, and B.L. Barkett. 2000. Variability of bed mobility in natural, gravel-bed channels and adjustments to sediment load at local and reach scales. Water Resour. Res. 36(12):3743-3756.

Loar, J.M., M.J. Sale, G.F. Cada, D.K. Cox, R.M. Cushman, G.K. Eddlemon, J.L. Elmore, A.J. Gatz, P. Kanciruk, J.A. Solomon, and D.S. Vaughan. 1985. Application of Habitat Evaluation Models in Southern Appalachian Trout Streams. ORNL/TM-9323. Environmental Sciences Division Publication No. 2383. Oak Ridge, TN: Oak Ridge National Laboratory. January 1985.

Lobb, M.D., III, and D. J. Orth. 1991. Habitat use by an assemblage of fish in a large warmwater stream. Trans. Am. Fish. Soc. 120(1):65-78.

Lund, J.R., and R.N. Palmer. 1997. Water resource system modeling for conflict resolution. Water Resources Update 108(Summer):70-82.

Manly, B.F.J., L.L. McDonald, and D.L. Thomas. 1993. Resource Selection by Animals: Statistical Design and Analysis for Field Studies, 1st Ed. New York: Chapman and Hall.

Manly, B.F.J., L.L. McDonald, D.L. Thomas, T.L. McDonald, and W.P. Erickson. 2002. Resource Selection by Animals: Statistical Design and Analysis for Field Studies, 2nd Ed. New York: Springer.

Marthur, D., W.H. Basson, E.J. Purdy, Jr., and C.A. Silver. 1985. A critique of the Instream Flow Incremental Methodology. Can. J. Fish. Aquat. Sci. 42(4):825-831.

Maslin, M.A., S. Burns, H. Erlenkeusser, and C. Hohnemann. 1997. Stable isotope records from ODP Sites 932 and 933. Pp. 305-318 in Proceedings of the Ocean Drilling Program, Vol. 155. Scientific Results, Amazon Fan, R.D. Flood, D.J.W. Piper, A. Klaus, and L.C. Peterson, eds. Ocean Drilling Program, Texas A&M University, College Station, TX [online]. Available: http://www-odp.tamu.edu/publications/155_SR/CHAP_17.PDF [accessed April 19, 2007].

Matanga, G.B., K.E. Nelson, E Sudicky, R. Therrien, S. Panday, R. McLaren, D. Demarco, and L. Gessford. 2004. HydroSphere: Fully-Integrated, Surface/Subsurface Numerical Model for Watershed Analysis of Hydrologic, Water Quality, and Sedimentation Processes. Abstract H32B-03. American Geophysical Union Fall Annual Meeting, Washington, DC [online]. Available: http://adsabs.harvard.edu/abs/2004AGUFM.H32B.03M [accessed Oct 25, 2007].

McComick, S.D., and R.L. Saunders. 1987. Prepatory physiological adaptations for marine life of salmonids: Osmoregulation, growth, and metabolism. Pp. 211-229 in Common Strategies of Anadromous and Catadromous Fishes, M.J. Dadswell, ed. American Fisheries Society Symposium No. 1. Bethesda, MD: American Fisheries Society.

McLean, S.R., S.R. Wolfe, and J.M. Nelson. 1999. Predicting boundry shear stress and sediment transport over bed forms. J. Hydraul. Eng. 125(7):725-736.

Milhous, R.T. 1982. Effect of sediment transport and flow regulation on the ecology of gravel-bed rivers. Pp. 819-842 in Gravel-Bed Rivers: Fluvial Processes, Engineering, and Management: Proceedings of International Workshop on Engineering Problems in the Management of Gravel-Bed Rivers, R.D. Hay, J.C. Bathust, and C.R. Thorne, eds. Chichester: Wiley.

Milhous, R.T., D.L. Wegner, and T. Waddle. 1984. User's Guide to the Physical Habitat Simulation System (PHABSIM). FWS/OBS-81/43 Revised. Instream Flow Information Paper 11. U.S. Fish and Wildlife Service, U.S. Department of the Interior, Washington, DC.

Milhous, R.T., M.A. Updike, and D.M. Schneider. 1989. Physical Habitat Simulation System Reference Manual—Version II. Instream Flow Information Paper No. 26. Biological Report 89(16). U.S. Fish and Wildlife Service, U.S. Department of the Interior, Washington, DC [online]. Available: http://www.mesc.usgs.gov/products/publications/3912/3912.pdf [accessed Feb. 15, 2007].

Milhous, R.T., J.M. Bartholow, M.A. Updike, and A.R. Moos. 1990. Reference Manual for the Generation and Analysis of Habitat Time Series—Version II. Instream Flow Information Paper No. 27. Biological Report 90(16). U.S. Fish and Wildlife Service, U.S. Department of the Interior, Washington, DC.

Monteith, J.L. 1965. Evaporation and environment. Pp. 205-233 in The State and Movement of Water in Living Organisms: Proceedings of the 19th Symposium of the Society of Experimental Biology, G.E. Fogg, ed. New York, NY: Cambridge University Press.

Mount, J.F. 1995. California Rivers and Streams: the Conflict between Fluvial Processes and Land Use. Berkeley, CA: University of California Press.

Moyle, P.B. 2002. Inland Fishes of California. Berkeley, CA: University of California Press. 502 pp.

Moyle, P.B., M.P. Marchetti, J. Baldrige, and T.L. Taylor. 1998. Fish health and diversity: Justifying flows for a California stream. Fisheries 23(7):6-15.

Myrick, C.A., and J.J. Cech, Jr. 2000. Swimming performances of four California stream fishes: Temperature effects. Environ. Biol. Fish. 58(3):289-295.

NCEP (National Centers for Environmental Prediction). 2006. The NCEP 51-year Hydrological Reanalysis Archive (also known as NLDAS data). National Center for Atmospheric Research and the University Corporation for Atmospheric Research http://dss.ucar.edu/pub/hydro51/ [accessed April 12, 2007].

NCEP (National Centers for Environmental Prediction). 2007. North American Regional Reanalysis [online]. Available: http://www.emc.ncep.noaa.gov/mmb/rreanl/ [accessed April 12, 2007].

NCRWQB (North Coast Regional Water Quality Control Board). 2006. RMA-2 and RMA-11 Flow and Water Quality Modeling Results for Natural Flows and Existing Conditions within the Klamath River for TDML Assessments (as cited in Hardy et al. 2006).

Nelson, J.M. 1996. Predictive techniques for river channel evolution and maintenance. Water Air Soil Poll. 90(1-2):321-333.

Nelson, J.M., R.L. Shreve, R. McLean, and T.G. Drake. 1995. Role of near-bed turbulence structure in bed load transport and bed form mechanics. Water Resour. Res. 31(8):2071-2086.

NMFS (National Marine Fisheries Service). 2001. Biological Opinion. Ongoing Klamath Project Operations. National Marine Fisheries Service, Southwest Region, National Oceanic and Atmospheric Administration, Long Beach, CA. April 6, 2001 [online]. Available: http://www.usbr.gov/mp/kbao/esa/38_cohobo_4_6-01.pdf [accessed April 5, 2007].

NMFS (National Marine Fisheries Service). 2002. Biological Opinion, Klamath Project Operations. National Marine Fisheries Service. May 31, 2002 [online]. Available: http://swr.nmfs.noaa.gov/psd/klamath/KpopBO2002finalMay31.pdf [accessed April 5, 2007].

NOAA (National Oceanic and Atmospheric Administration). 1972. National Weather Service River Forecast System Forecast Procedures. NOAA Tech. Memo. NWS Hydro-14. Office of Hydrologic Development, Hydrology Laboratory, National Weather Service, National Oceanic and Atmospheric Administration, Silver Spring, MD. December 1, 1972.

NRC (National Research Council). 1990. Ground Water Models: Scientific and Regulatory Applications. Washington, DC: National Academy Press.

NRC (National Research Council). 1994. Review of EPA's Environmental Assessment and Monitoring Program: Surface Waters. Washington, DC: National Academy Press

NRC (National Research Council). 1996. Upstream: Salmon and Society in the Pacific Northwest. Washington, DC: National Academy Press.

NRC (National Research Council). 2000. Ecological Indicators for the Nation. Washington, DC: National Academy Press.

NRC (National Research Council). 2002. Scientific Evaluation of Biological Opinions on Endangered and Threatened Fishes in the Klamath River Basin: Interim Report. Washington, DC: National Academy Press.

NRC (National Research Council). 2003. Adaptive Monitoring and Assessment for the Comprehensive Everglades Restoration Plan. Washington, DC: The National Academies Press.

NRC (National Research Council). 2004a. Endangered and Threatened Fishes in the Klamath River Basin: Causes of Decline and Strategies for Recovery. Washington, DC: The National Academies Press.

NRC (National Research Council). 2004b. Atlantic Salmon in Maine. Washington DC: The National Academies Press.

NRC (National Research Council). 2005a. The Science of Instream Flows: A Review of Texas Instream Flow Program. Washington, DC: The National Academies Press.

NRC (National Research Council). 2005b. Re-Engineering Water Storage in the Everglades: Risks and Opportunities. Washington, DC: The National Academies Press.

NRC (National Research Council). 2005c. Endangered and Threatened Species of the Platte River. Washington, DC: The National Academies Press.

NRC (National Research Council). 2006. Review of the Lake Ontario-St. Lawrence River Studies. Washington, DC: The National Academies Press.

NRC (National Research Council). 2007a. Models in Environmental Regulatory Decision Making. Washington, DC: The National Academies Press.

NRC (National Research Council). 2007b. Progress Toward Restoring the Everglades: The First Biennial Review, 2006. Washington, DC: The National Academies Press.

NRCS (Natural Resources Conservation Service). 2006. Klamath River Basin Overview. U.S. Department of Agriculture, Natural Resources Conservation Service [online]. Available: http://www.nrcs.usda.gov/feature/klamath/klambasin.html [accessed Dec. 29, 2006].

Ogden, J.C., S.M. Davis, K.J. Jacobs, T. Barnes, and H.E. Fling. 2005. The use of conceptual ecological models to guide ecosystem restoration in south Florida. Wetlands 25(4):795–809.

Oregon Lakes Association. 2005. Klamath Lake [online]. Available: http://www.oregonlakes.org/gallery/klamath/klamath.html [accessed Dec. 29, 2006].

Oreskes, N. 2003. The role of quantitative models in science. Pp. 13-31 in Models in Ecosystem Science, C.D. Canham, J.J. Cole, and W.K. Lauenroth, eds. Princeton: Princeton University Press.

Orth, D.J. 1987. Ecological considerations in the development and application of instream flow-habitat models. Regul. River. 1(2):171-181.

Osborne, L.L., M.J. Wiley, and R.W. Latimore. 1988. Assessment of the water surface profile model: Accuracy of predicted instream fish habitat conditions in low-gradient, warmwater streams. Regul. River. 2(5):619-631.

PacifiCorp. 2004. Final Technical Report-Klamath Hydroelectric Project (FERC Project No. 2082) Water Resources. PacifiCorp, Portland, OR. February 2004 [online]. Available: http://www.pacificorp.com/Article/Article35437.html [accessed April 20, 2007]

Palmer, R.N., W.J. Werick, A. MacEwan, and A.W. Woods. 1999. Modeling water resources opportunities, challenges and trade-offs: The use of shared vision modeling for negotiation and conflict resolution. In WRPMD 1999: Preparing for the 21st Century, 29th Annual Water Resources Planning and Management Conference, June 6–9, 1999, Tempe, AZ, E.M. Wilson, ed. American Society of Civil Engineers [online]. Available: http://www.tag.washington.edu/papers/papers/Palmer-etal.1999.ASCE-Conf-Proc.0-7844-0430-5.pdf [accessed Feb. 16, 2007].

Parker, P., R. Letcher, A. Jakeman, M.B. Beck, G. Harris, R.M. Argent, M. Hare, C. Pahl-Wostl, A. Voinov, M. Janssen, P. Sullivan, M. Scoccimarro, A. Friend, M. Sonnenshein, D. Barker, L. Matejicek, D. Odulaja, P. Deadman, K. Lim, G. Larocque, P. Tarikhi, C. Fletcher, A. Put, T. Maxwell, A. Charles, H. Breeze, N. Nakatani, S. Mudgal, W. Naito, O. Osidele, I. Eriksson, U. Kautsky, E. Kautsky, B. Naeslund, L. Kumblad, R. Park, S. Maltagliati, P. Girardin, A. Rizzoli, D. Mauriello, R. Hoch, D. Pelletier, J. Reilly, R. Olafsdottir, and S. Bin. 2002. Progress in integrated assessment and modeling. Environ. Model. Softw. 17(3):209–217.

Perry, T. 2006. Overview of USBR Report on Natural Flow of the Upper Klamath River. Presentation at the First Meeting on Hydrology, Ecology, and Fishes of the Klamath River Basin, February 13-14, 2006, Sacramento, CA.

Perry, T., C. Albertson, T. Mull, and J. Rassmussen. 2002. Technical Memorandum to Area Manager, Klamath Project Area Office, from U.S. Bureau Technical Services Center, Denver, CO. November 6, 2002. 14 pp.

Petit, F. 1987. The relationship between shear stress and the shaping of the bed of a pebble-loaded river La Rulles-Ardenne. Catena 14(5):453-468.

Poff, N.L., J.D. Allan, M.B. Bain, J.R. Karr, K.L. Prestegaard, B.D. Richter, R.E. Sparks, and J.C. Stromberg. 1997. The natural flow regime: A paradigm for river conservation and restoration. Bioscience 47(11):769-784.

Polya, G. 1957. How to Solve It: A New Aspect of Mathematical Method, 2nd Ed. Princeton: Princeton University Press.

PWA (Phillip Williams and Associates, Ltd.). 2002. Hydrodynamic Modeling of Upper Klamath Lake. Prepared for the U.S. Bureau of Reclamation, by Phillip Williams and Associates, Ltd. November 11, 2002. 17 pp.

Quade, E.S. 1980. Pitfalls in formulation and modeling. Pp. 23-43 in Pitfalls of Analysis, G. Majone, and E.S. Quade, eds. International Series on Applied Systems Analysis No. 8. New York: Wiley.

Railsback, S.F., H.B. Stauffer, and B.C. Harvey. 2003. What can habitat preference models tell us? Tests using a virtual trout population. Ecol. Appl. 13(6):1580-1594.

Rankel, G.L. 1982. An appraisal of the status and future of wild Chinook salmon in the Klamath River drainage. In Wild Trout and Steelhead Fisheries: Laws and Law Enforcement-Management of Wild Salmonids, Proceedings of a Symposium, K.A. Hashagen, ed. San Francisco: California Trout Inc. (as cited in Hardy et al. 2006).

Richey, J. 2006. Recovery of Fall-Run Chinook and Coho Salmon at Iron Gate Hatchery. California Department of Fish and Game, Klamath River Project, Yreka, CA.

Risley, J.C., and A. Laenen. 1999. Upper Klamath Lake Basin Nutrient Loading Study-Assessment of Historic Flows in the Williamson and Sprague Rivers. Survey Water Resources Investigation Report 98-4198. Portland, OR: U.S. Department of the Interior, U.S. Geological Survey [online]. Available: http://or.water.usgs.gov/pubs_dir/Pdf/98-4198.pdf [accessed April 5, 2007].

Risley, J.C., M.W. Gannett, J.K. Lea, and E.A. Roehl, Jr. 2005. An Analysis of Statistical Methods for Seasonal Flow Forecasting in the Upper Klamath River Basin of Oregon and California. Scientific Investigations Report 2005-5177. U.S. Department of the Interior, U.S. Geological Survey [online]. Available: http://pubs.usgs.gov/sir/2005/5177/pdf/sir2005-5177.pdf [accessed Feb. 22, 2007].

Ropella, G.E., S.F. Railsback, and S.K. Jackson. 2002. Software engineering considerations for individual-based models. Nat. Resour. Model. 15(1):5-22.

Rose, K.A. 1989. Sensitivity analysis in ecological simulation models. Pp. 4230-4234 in Systems and Control Encyclopedia, M.G. Singh, ed. New York: Pergamon Press.

Salas, J.D. 1993. Analysis and modeling of hydrologic time series. Pp. 19.1-19.72 in Handbook of Hydrology, D.R. Maidment, ed. New York: McGraw-Hill.

Saltelli, A., K. Chan, and E.M. Scott, eds. 2000. Sensitivity Analysis. New York: Wiley.

Satkowski, R., J.R. Lund, H. Morel-Seytoux, A. Nelson, and T. Roefs. 2000. Protocols for Water and Environmental Modeling. Report No. BDMF 2000-1. Bay-Delta Modeling Forum, Richmond, CA. January 21, 2000 [online]. Available: http://www.cwemf.org/Pubs/Protocols2000-01.pdf [accessed Feb. 16, 2007].

Schleusner, D. 2006. Organizational Dynamics in the Trinity River Restoration Program. Presentation at AWRA 2006 Summer Specialty Conference: Adaptive Management of Water Resources, June 26-28, 2006, Missoula, MT.

Schlosser, I.J., and P.L. Angermeier. 1995. Spatial variation in demographic processes of lotic fishes: Conceptual models, empirical evidence, and implications for conservation. Pp. 392-401 in Evolution and the Aquatic Ecosystem: Defining Unique Units in Population

Conservation, J.L. Nielsen, ed. American Fisheries Society Symposium 17. Bethesda, MD: American Fisheries Society.

Schmidt, L.J., and J.P. Potyondy. 2004. Quantifying Channel Maintenance Instream Flows: An Approach for Gravel-Bed Streams in the Western United States. General Technical Report RMRS-GTR-128. U.S. Department of Agriculture, Forest Service, Rocky Mountain Research Station, Fort Collins, CO [online]. Available: http://www.fs.fed.us/rm/pubs/rmrs_gtr128.pdf [accessed April 20, 2007].

Scott, J.M., P.J. Heglund, M.L. Morrison, J.B. Haufler, M.G. Raphael, W.A. Wall, and F.B. Samson, eds. 2002. Predicting Species Occurrences: Issues of Accuracy and Scale. Washington, DC: Island Press.

SCS (Soil Conservation Service). 1970. Irrigation Water Requirements. Technical Release No. 21. U.S. Department of Agriculture, Soil Conservation Service, Engineering Division, Washington DC.

Serreze, M.C., M.P. Clark, R.L. Armstrong, D.A. McGinnis, and R.S. Pulwarty. 1999. Characteristics of the western United States snowpack from snowpack telemetry (SNOTEL) data. Water Resour. Res. 35(7):2145-2160.

Servizi, J.A., and D.W. Martens. 1991. Effect of temperature, season, and fish size on acute lethality of suspended sediments to coho salmon (*Oncorhynchus kisutch*). Can. J. Fish. Aquat. Sci. 48(3):493-497.

Servizi, J.A., and D.W. Martens. 1992. Sublethal responses of coho salmon (*Oncorhynchus kisutch*) to suspended sediments. Can. J. Fish. Aquat. Sci. 49(7):1389-1395.

Shaw Historical Library. 1999. A River Never the Same: A History of Water in the Klamath Basin. J. Shaw Hist. Lib. 13.

Shaw Historical Library. 2002. And Then We Logged—The Timber Industry in the Klamath Basin. J. Shaw Hist. Lib. 16.

Shirvell, C.S. 1986. Pitfalls of Physical Habitat Simulation in the Instream Flow Incremental Methodology. Canadian Technical Report of Fisheries and Aquatic Sciences No. 1460. Prince Rupert, BC: Fisheries and Oceans, Canada.

Shirvell, C.S. 1994. Effect of changes in streamflow on the microhabitat use and movements of sympatric juvenile coho salmon (*Oncorhynchus kisutch*) and chinook salmon (*O. tshawystcha*) in a natural stream. Can. J. Fish. Aquat. Sci. 51(7):1644-1652.

Singh, V.P., ed. 1995. Computer Models of Watershed Hydrology. Highlands Ranch, CO: Water Resources Publications.

Singh, V.P., and D.K. Frevert. 2006. Watershed Models. Boca Raton, FL: CRC Press.

Sobey, R.J. 2001. Standardized evaluation of estuarine network models. Pp. 3683-3696 in Coastal Engineering 2000: Conference Proceedings: 27th International Conference on Coastal Engineering, July 16-21, 2000, Sydney, Australia, B.L. Edge, ed. Reston, VA: American Society of Civil Engineers.

Sorooshian, S., and V.K. Gupta. 1995. Model calibration. Pp. 23-63 in Computer Models of Watershed Hydrology, V.P. Singh, ed. Highlands Ranch, CO: Water Resources Publications.

Stalnaker, C.B. 1979. The use of habitat structure preferrenda for establishing flow regimes necessary for maintenance of fish habitat. Pp. 321-337 in The Ecology of Regulated Streams, J.V. Ward and J.A. Stanford, eds. New York: Plenum Press.

Stalnaker, C., B.L. Lamb, J. Henriksen, K. Bovee and J. Bartholow. 1995. The Instream Flow Incremental Methodology—A Primer for IFM. Biological Report 29. U.S. Department of the Interior, National Biological Service, Washington DC [online]. Available: http://www.krisweb.com/biblio/gen_nbs_stalnakeretal_1995_ifim.pdf [accessed April 19, 2007].

Stewart, D.J., and M. Ibarra. 1991. Predation and production by salmonine fishes in Lake Michigan, 1978-88. Can. J. Fish. Aquat. Sci. 48(5):909-922.

Stockholm University. 2003. Penman's Equation. Penmanetc. Biogeochemical Modeling Node of the Land-Ocean Interactions in the Costal Zone Project of the International Geosphere Programme of the International Council of Scientific Unions [online]. Available: http://data.ecology.su.se/MNODE/Methods/penman.htm [accessed June 5, 2003].

Stocking, R.W., and J.L. Bartholomew. 2004. Assessing Links Between Water Quality, River Health and Ceratomyxosis of Salmonids in the Klamath River System. Department of Microbiology, Oregon State University, Corvallis, OR. 5 pp [online]. Available: http://www.klamathwaterquality.com/klamath_cShasta_Stocking_Bartholomew_2004.pdf [accessed June 26, 2007].

Stonecypher, R.W., W.A. Hubert, and W.A. Gern. 1994. Effect of reduced incubation temperatures on survival of trout embryos. Prog. Fish Cult. 56(3):180-184.

Stutzer, G.M., J. Ogawa, N.J. Hetrick, and T. Shaw. 2006. An Initial Assessment of Radiotelemetry for Estimating Juvenile Coho Salmon Survival, Migration Behavior, and Habitat Use in Response to Iron Gate Dam Discharge on the Klamath River, California. Arcata Technical Report No TR2006-05. U. S. Fish and Wildlife Service, Arcata Fish and Wildlife Office, Arcata, CA.

Tanaka, S.K., M.L. Deas, R.J. Sutton, T. Soto, and A. Corum. 2006. Physical Characterization of a Klamath River Thermal Refugia under Various Hydrologic and Meteorological Conditions. Draft report. 33 pp.

Tarboton, D.G., A. Sharma, and U. Lall. 1998. Disaggregation procedures for stochastic hydrology based on nonparametric density estimation. Water Resour. Res. 34(1):107-119.

Tesfaye, Y.G., M.M. Meerschaert, and P.L. Anderson. 2006. Identification of periodic autoregressive moving average models and their application to the modeling of river flows. Water Resour. Res. 42, W01419, doi:10.1029/2004WR003772.

Tetra Tech, Inc. 2007. Klamath River Model for TMDL Development. Prepared for U.S. Environmental Protection Agency Region 9 and 10, Oregon Department of Environmental Quality, North Coast Regional Water Quality Control Board. September 2007.

Therrien, R., R.G. McLaren, E.A. Sudicky, and S. Panday. 2004. HydroSphere: A Three-Dimensional Numerical Model Describing Fully-Integrated Subsurface and Surface Flow and Solute Transport, User's Guide. Université Laval and University of Waterloo, Waterloo, ON. 273 pp.

Thomas, J.A., and K.D. Bovee. 1993. Application and testing of a procedure to evaluate transferability of habitat suitability criteria. Regul. River. 8(3):285-294.

Thompson, D.M., J.M. Nelson, and E.E. Wohl. 1998. Interactions between pool geometry and hydraulics. Water Resour. Res. 34(12):3673-3681.

Topping, D.J., D.M. Rubin, J.M. Nelson, P.J. Kinzel III, and I.C. Corson. 2000. Colorado River sediment transport. 2. Systematic bed-elevation and grain-size effects of sand supply limitation. Water Resour. Res. 36(2):543-570.

Trinity Management Council Subcommittee. 2004. Trinity River Restoration Program Evaluation. Final Report. March 2004.

Trush, W.J., S.M. McBain, and L.B. Leopold. 2000. Attributes of an alluvial river and their relation to water policy and management. Proc. Natl. Acad. Sci. U.S.A. 97(22): 11858-11863.

USACE (U.S. Army Corps of Engineers). 2007. HEC-RAS and HEC-HMS Software. The Hydrologic Engineering Center, U.S. Army Corps of Engineers [online]. Available: http://www.hec.usace.army.mil/ [accessed April 5, 2007].

USBR (U.S. Bureau of Reclamation). 2000. Klamath Project Historic Operation. U.S. Department of the Interior, Bureau of Reclamation, Mid-Pacific Region, Klamath Basin Area Office. November 2000 [online]. Available: http://www.usbr.gov/mp/kbao/docs/Historic%20Operation.pdf [accessed Feb. 22, 2007].

USBR (U.S. Bureau of Reclamation). 2001a. Biological Assessment of Klamath Project's Continuing Operations on the Endangered Lost River Sucker and Shortnose Sucker. U.S.

Department of the Interior, Bureau of Reclamation, Mid-Pacific Region, Klamath Basin Area Office, Klamath Falls, OR. February 13, 2001 [online]. Available: http://www. mp.usbr.gov/kbao/esa/34_final_sucker_bo_4_06_01.pdf [accessed April 5, 2007].

USBR (U.S. Bureau of Reclamation). 2001b. Biological Assessment of the Klamath Project's Continuing Operations on Southern Oregon/Northern California ESU Coho Salmon and Critical Habitat for Southern Oregon/Northern California ESU Coho Salmon. U.S. Department of the Interior, Bureau of Reclamation, Mid-Pacific Region, Klamath Basin Area Office, Klamath Falls, OR. January 22, 2001 [online]. Available: http://www.usbr. gov/mp/kbao/esa/Final_Ba_012201_Sutton.pdf [accessed April 5, 2007].

USBR (U.S. Bureau of Reclamation). 2002. Final Biological Assessment—The Effects of Proposed Actions Related to Klamath Project Operation (April 1, 2002-March 31, 2012) on Federally Listed Threatened and Endangered Species. U.S. Department of the Interior, Bureau of Reclamation, Mid-Pacific Region, Klamath Basin Area Office. February 25, 2002 [online]. Available: http://www.usbr.gov/mp/kbao/docs/Final_Biological_Assessment_02-25-02.pdf [accessed Feb. 22, 2007].

USBR (U.S. Bureau of Reclamation). 2005. Natural Flow of the Upper Klamath River-Phase I. Prepared by Technical Service Center, Denver, CO, for U.S. Department of the Interior, Bureau of Reclamation, Klamath Basin Area Office, Klamath Falls, OR. November 2005 [online]. Available: http://www.onebasin.org/documents/200511naturalflow.pdf [accessed Feb. 20, 2007].

USBR (U.S. Bureau of Reclamation). 2006. Klamath Project 2006 Operations Plan. April 10, 2006 [online]. Available: http://www.klamathbasincrisis.org/BOR/2006_Klamath_Project_Operations_Plan.pdf [accessed Dec. 13, 2007].

USBR (U.S. Bureau of Reclamation). 2007a. AgriMet: The Pacific Northwest Cooperative Agricultural Weather Network [online]. Available: http://www.usbr.gov/pn/agrimet/webarcread.html? [accessed April 10, 2007].

USBR (U.S. Bureau of Reclamation). 2007b. Conservation Implementation Program (CIP). U.S. Bureau of Reclamation [online]. Available: http://www.usbr.gov/mp/kbao/CIP/index. html [accessed April 30, 2007].

USFWS (U.S. Fish and Wildlife Service). 2001. Biological/Conference Opinion Regarding the Effects of Operation of the Bureau of Reclamation's Klamath Project on the Endangered Lost River Sucker (*Deltistes luxatus*), Endangered Shortnose Sucker (*Chasmistes brevirostris*), Threatened Bald Eagle (*Haliaeetus leucocephalus*), and Proposed Critical Habitat for the Lost River/Shortnose Suckers. U.S. Fish and Wildlife Service, Klamath Falls Fish and Wildlife Office, Klamath Falls, OR. April 2001 [online]. Available: http://www.usbr. gov/mp/kbao/esa/34_final_sucker_bo_4_06_01.pdf [accessed April 5, 2007].

USFWS (U.S. Fish and Wildlife Service). 2002. Biological/Conference Opinion Regarding the Effects of Operation of the U.S. Bureau of Reclamation's Proposed 10-Year Operation Plan for the Klamath Project and its Effect on the Endangered Lost River Sucker (*Deltistes luxatus*), Endangered Shortnose Sucker (*Chasmistes brevirostris*), Threatened Bald Eagle (*Haliaeetus leucocephalus*), and Proposed Critical Habitat for the Lost River and Shortnose Suckers. U.S. Fish and Wildlife Service, Klamath Falls Fish and Wildlife Office, Klamath Falls, OR.

USFWS/HVT (U.S. Fish and Wildlife Service and Hoopa Valley Tribe). 1999. Trinity River Flow Evaluation. Final Report. U.S. Fish and Wildlife Service, Arcata Fish and Wildlife Office, Arcata, CA, and Hoopa Valley Tribe, Hopa CA. June 1999 [online]. Available: http://www.trrp.net/documents/Trinity_River_Flow_Evaluation_Final_Report.pdf [accessed March, 2007].

USGS (U.S. Geological Survey). 2005. Assessment of the Klamath Project Pilot Water Bank: A Review from a Hydrologic Perspective. Prepared for U.S. Bureau of Reclamation, Klamath Basin Area Office, Klamath Falls, OR, by U.S. Geological Survey, Oregon Water

Science Center, Portland, OR [online]. Available: http://www.usbr.gov/mp/kbao/docs/Final_USGS_Assessment_of_Water_Bank.pdf [accessed Feb. 22, 2007].

USGS (U.S. Geological Survey). 2007a. National Water Information System: Web Interface [online]. Available: http://waterdata.usgs.gov/usa/nwis/ [accessed Oct. 29, 2007].

USGS (U.S. Geological Survey). 2007b. The Five Phases of IFIM. Instream Flow Incremental Methodology [online]. Available: http://www.fort.usgs.gov/products/software/ifim/ [accessed Oct. 30, 2007].

Van Cleve, F.B., C. Simenstad, F. Goetz, and T. Mumford. 2004. Application of the "Best Available Science" in Ecosystem Restoration: Lessons Learned from Large-Scale Restoration Project Efforts in the USA. Technical Report 2004-01. Puget Sound Nearshore Partnership [online]. Available: http://pugetsoundnearshore.org/techical_papers/lessons.pdf [accessed Oct. 9, 2007].

Vannote, R.L., G.W. Minshall, K.W. Cummins, J.R. Sedell, and E. Gushing. 1980. The river continuum concept. Can. J. Fish. Aquat. Sci. 37(1):130-137.

Waddle, T., P. Steffler, A. Ghanem, C. Katapolis, and A. Locke. 2000. Comparison of one- and two-dimensional open channel flow models for a small habitat stream. Rivers 7(3):205-220.

Waddle, T.J., ed. 2001. PHABSIM for Windows: User's Manual and Exercises. Open File Report 01-340. Fort Collins, CO: U.S. Geological Survey. 288 pp. [online]. Available: http://www.fort.usgs.gov/products/publications/15000/15000.pdf [accessed Feb. 22, 2007].

Wagener, T., H.S. Wheater, and H.V. Gupta. 2004. Rainfall–Runoff Modelling in Gauged and Ungauged Catchments. London, UK: Imperial College Press.

Walters, C., J. Korman, L.E. Stevens, and B. Gold. 2000. Ecosystem modeling for evaluation of adaptive management policies in the Grand Canyon. Conserv. Ecol. 4(2):1 [online]. Available: http://www.ecologyandsociety.org/vol4/iss2/art1/ [accessed Feb. 22, 2007].

Ward, J.V. 1989. The four-dimensional nature of lotic ecosystems. J. N. Am. Benthol. Soc. 8(1):2–8.

Waters, T. 1995. Sediment in Streams: Sources, Biological Effects, and Control. American Fisheries Society Monograph 7. Bethesda, MD: American Fisheries Society.

Weather Underground. 2007. History for Klamath Falls, OR [online]. Available: http://www.wunderground.com/history/airport/KLMT/2006/12/31/DailyHistory.html?req_city=NA&req_state=NA&req_statename=NA [accessed April 23, 2007].

Weddell, B.J. 2000. Relationship between Flows in the Klamath River and Lower Klamath Lake Prior to 1910. Prepared for the U.S. Department of the Interior, Fish and Wildlife Service, Klamath Basin Refuges, Tule Lake, CA. November 28, 2000.

Weitkamp, L.A., T.C. Wainwright, G.J. Bryant, G.B. Milner, D.J. Teel, R.G. Kope, and R.S. Waples. 1995. Status Review of Coho Salmon from Washington, Oregon, and California. NOAA Technical Memorandum NMFS-NWFSC-24. Seattle, WA: U.S. Department of Commerce, National Oceanic and Atmospheric Administration, National Marine Fisheries Service, Northwest Fisheries Science Center. 258 pp [online]. Available: http://www.nwfsc.noaa.gov/publications/techmemos/tm24/tm24.htm [accessed April 10, 2007].

Wells, S.A., R. Annear, and M. McKillip. 2004. Review of the Klamath River Model for Klamath Hydropower Project FERC 32082. Draft. Prepared for The Bureau of Land Management and Karuk Tribe. May 3, 2004 [online]. Available: http://www.klamathwaterquality.com/klamath_wells_draft_fla_model_2004.pdf [accessed Oct. 30, 2007].

Whiting, P.J. 1997. The effect of stage on flow and components of the local force balance. Earth Surf. Proc. Land. 22(6):517-530.

Whiting, P.J., and W.E. Dietrich. 1991. Convective accelerations and boundary shear stress over a channel bar. Water Resour. Res. 27(5):783-796.

WMO (World Meteorological Organization). 1975. Intercomparison of Conceptual Models

Used in Operational Hydrological Forecasting. Operational Hydrology Report No. 7. WMO Publ. No. 429. Geneva: World Meteorological Organization. 172 pp.

Williams, J.G. 1996. Lost in space: Minimum confidence intervals for idealized PHABSIM studies. Trans. Am. Fish. Soc. 125(3):458-465.

Williams, J.G. 1999. Stock dynamics and adaptive management of habitat: An evaluation based on simulations. N. Am. J. Fish. Manage. 19(2):329-341.

Williamson, S.C., J.M. Bartholow, and C.B. Stalnaker. 1993. Conceptual model for quantifying pre-smolt production from flow-dependent physical habitat and water temperature. Regul. River. 8(1-2):15-28.

Woodling, J.D. 1994. The South Platte River from Denver to Nebraska: Monitoring Water Quality Is Not a Simple Process. Pp. 46-47 in Integrated Watershed Management in the South Platte Basin: Status and Practical Implementation, Proceedings of the 1994 South Platte Forum, October 26-27, 1994, Greeley, CO, K.C. Klein, ed. Information Series No. 77. Colorado Water Resources Research Institute, Colorado State University, Fort Collins, CO [online]. Available: http://cwrri.colostate.edu/pubs/series/information/IS77.pdf [accessed April 26, 2007].

Wurbs, R.A., and E.D. Sisson. 1999. Comparative Evaluation of Methods for Distributing Naturalized Streamflows from Gauge to Ungauged Sites. Technical Report No. 179. College Station, TX: Texas Water Resources Institute, The Texas A&M University System [online]. Available: http://borderwater.tamu.edu/reports/1999/179/tr179.pdf [accessed April 18, 2007].

Yeh, G.T., G. Huang, H.P. Cheng, F. Zhang, H.C. Lin, E. Edris, and D. Richards. 2006. A First principle, physics-based watershed model: WASH123D. Pp. 211-244 in Watershed Models, V.P. Singh, and D. Frevert, eds. New York: Taylor and Francis Group, CRC Press.

Yreka Chamber of Commerce. 2007. Yreka, California: On the National Register of Historical Places, 2006 Economic Profile [online]. Available: www.yrekachamber.com/econ.html [accessed April 10, 2007].

Appendix

Biographical Information for Committee Members

William L. Graf (*Chair*) is Foundation University Professor and professor of geography at the University of South Carolina. His specialties include fluvial geomorphology and hydrology, as well as policy for public land and water. His Ph.D. is from the University of Wisconsin, Madison, with a major in physical geography and a minor in water resources management. His research and teaching have focused on river-channel change and human impacts on river processes, including the downstream effects of large dams. He has authored or edited 9 books, more than 130 scientific papers, book chapters, and reports, more than 60 successful grant proposals, and more than 100 public presentations. He is past president of the Association of American Geographers and is a National Associate of the National Academy of Science. He has chaired numerous National Research Council committees dealing with river science and policy. He worked with the Presidential Commission on Western Water and was appointed to the Presidential Commission on American Heritage Rivers.

Michael E. Campana directs the Institute for Water and Watersheds at Oregon State University, where he is also professor of geosciences. He was formerly the Albert J. and Mary Jane Black Professor of Hydrogeology and director of the Water Resources Program at the University of New Mexico, where he worked from 1989-2006. He was a research hydrologist at the Desert Research Institute in Reno and taught in the University of Nevada's Hydrologic Sciences Program from 1976 to 1989. His interests include hydrophilanthropy, water resources in developing countries, transboundary water resources issues, regional hydrogeology, surface water-ground water interactions, and arid zone hydrology. He has supervised the work of 66 graduate students and has authored/co-authored over 70 reports and journal articles. His international work is primarily in Nicaragua, Honduras, Panama, Kazakhstan, and the South Caucasus, where he directs the six-country NATO/OSCE project South Caucasus River Monitoring. He was a Fulbright Scholar to Belize in 1996 and a Visiting Scientist at the Research Institute for Groundwater in Cairo (Fall 1995) and the International Atomic Energy Agency in Vienna (Fall 2002). He has served on several previous National Research Council committees (USGS research, NAWQA Program) and currently serves on the Sustainable Oceans, Coasts and Waterways Advisory Committee of the H. J. Heinz Center. Dr. Campana is founder, president, and treasurer of the Ann Campana Judge Foundation (www.acjfoundation.org), a 501(c)(3) charitable foundation that funds and undertakes projects related to water, health, and sanitation in developing countries. He earned his BS in geology from the College of William and Mary and his MS and Ph.D. degrees in hydrology from the University of Arizona.

George Mathias Kondolf is a professor of environmental planning and geography at the University of California at Berkeley, where he teaches hydrology for planners, restoration of rivers and streams, ecological analysis in urban design, and introduction to environmental sciences. He earned an A.B. in geology (cum laude) from Princeton University, an M.S. in earth sciences from the University of California at Santa Cruz, and a Ph.D. in geography and environmental engineering from the Johns Hopkins University. He is a fluvial geomorphologist whose research concerns environmental river management, influences of land use on rivers, notably the effects of mining and dams on river systems, interactions of riparian vegetation and channel form, geomorphic influences on habitat for salmon and trout, alternative flood management strategies, and assessment of ecological restoration. In addition to numerous technical papers on these and related topics, he recently published the reference work *Tools in Fluvial Geomorphology* (Wiley 2003). Dr. Kondolf has served as a consultant to clients including the Federal Republic of Germany, the U.S. Fish and Wildlife Service, the U.S. Army Corps of Engineers, the U.S. Bureau of Land Management, the California Attorney General, the California Department of Fish and Game and Department of Water Resources, various water districts and utilities, aggregate producers, and environmental organizations. Dr. Kondolf is currently a member of the Environmental Advisory Board to the Chief of the U.S. Army Corps of Engineers.

Jay R. Lund is a professor of civil and environmental engineering at the University of California, Davis. He earned a B.A. in regional planning and international relations at the University of Delaware in 1979, a B.S. in civil engineering at the University of Washington in 1983, an M.A. in geography at the University of Washington in 1983, and a Ph.D. in civil engineering at the University of Washington in 1986. He served on advisory committees for the 1998 and 2005 California Water Plans, as convenor of the California Water and Environment Modeling Forum, and editor of the *Journal of Water Resources Planning and Management*. He is a member of the International Water Academy and has won several awards for water-related research from the American Society of Civil Engineers. He is the principal developer of the CALVIN economic-engineering optimization model of California's water supply system, applied regionally and statewide to explore water markets, conjunctive use, integrated water management, climate change, and environmental restoration. He has had a major role in water and environmental system modeling projects in California, the United States, and overseas. His principal specialties are simulation, optimization, and management of large-scale water and environmental systems, the application of economic ideas and methods, reservoir operation theory, and water demand theory and methods. He is author or co-author of over 180 publications.

Judith L. Meyer is a distinguished research professor emeritus at the University of Georgia's Institute of Ecology and former director of the River Basin Center. She currently serves on the NRC Board of Environmental Studies and Toxicology and has served on the Water Science and Technology Board and several NRC committees. She is a past president of the Ecological Society of America. She currently chairs the Environmental Processes and Effects Committee of the EPA's Science Advisory Board. She chairs the Scientific and Technical Advisory Committees of American Rivers and is vice-chair of the Independent Science Board of the California Bay Delta Authority. Her expertise is in river and stream ecosystems with emphasis on nutrient dynamics, microbial food webs, riparian zones, ecosystem management, river restoration, and urban rivers. She received the 2003 Award of Excellence in Benthic Science from the North American Benthological Society. Dr. Meyer earned a B.S. from University of Michigan, an M.S. from University of Hawaii, and a Ph.D. from Cornell University.

Dennis D. Murphy is research professor in the Biology Department and director of the graduate program in ecology, evolution, and conservation biology at the University of Nevada, Reno. He earned a B.S. at the University of California, Berkeley and a Ph.D. from Stanford University. Until recently, he served as director and then as President of the Center for Conservation Biology at Stanford. Author of more than 170 published papers and book chapters on the biology of butterflies and on key issues in the conservation of imperiled species, Dr. Murphy has worked in conflict resolution in land-use planning on private property since the first federal Habitat Conservation Plan (HCP) on San Bruno Mountain, including HCPs in the Pacific Northwest, southern California, and Nevada. He won the industry's oldest and most respected prize in conservation, the Chevron Conservation Award; has been named a Pew Scholar in Conservation and the Environment; and has received the California Governor's Leadership Award in Economics and the Environment. Dr. Murphy has served a number of scientific societies and environmental organizations and is past president of the Society for Conservation Biology. His professional activities outside academia include service on the Interagency Spotted Owl Scientific Advisory Committee, enjoined by Congress to develop a solution to that planning crisis in the Pacific Northwest, as chair of the National Park Service's Scientific Advisory Committee on Bighorn Sheep, as cochair of the Department of State's American-Russian Young Investigators Program in Biodiversity and Ecology, as codirector of the statewide Nevada Biodiversity Initiative based at the University of Nevada at Reno, and as chair of the Scientific Review Panel of the first Natural Community Conservation Planning Program in southern California's coastal sage scrub ecosystem. He served the National Research Council on its Committee on Endangered and Threatened Species in the Platte River Basin and Committee on Scientific Issues in the Endangered Species Act and in its contribution to the recent General Accounting Office review of desert tortoise management and recovery. He has been a member of both the Applied Science Panel and the Interagency Working Group of the federal-state Coastal Salmon Initiative in northern California.

Christopher A. Myrick is assistant professor of fishery biology and aquaculture in the Department of Fishery and Wildlife Biology at Colorado State University. Dr. Myrick is also facility manager of the Foothills Fisheries Laboratory of Colorado State University. He earned a B.S. in resource management at the University of California (Berkeley) and an M.S. and Ph.D. in ecology from the University of California (Davis). Dr. Myrick's research interests are in culture and physiology of native fishes, physiological ecology of fishes, and aquatic ecology. He has written numerous articles on trout, steelhead, and salmonids in California. Other research has explored the temperature effects on physiological mechanisms within fishes.

Tammy J. Newcomb is the Lake Huron Basin Coordinator for the Michigan Department of Natural Resources Fisheries Division. In this position, she coordinates ecosystem and watershed management for the Lake Huron drainages and the Lake Huron sport, tribal, and commercial fisheries. Dr. Newcomb is an adjunct faculty member at Virginia Polytechnic Institute and State University and Michigan State University. Her research focus is on salmonid population dynamics, watershed and stream habitat management, instream flow management, and stream temperature modeling. Dr. Newcomb earned a Ph.D. at Michigan State University. Dr. Newcomb served on the NRC Committee on Endangered and Threatened Fishes in the Klamath River Basin and the Committee on Water Resources Management, Instream Flows, and Salmon Survival in the Columbia River.

Jayantha Obeysekera earned a B.S. degree in civil engineering from University of Sri Lanka; an M.E. in hydrology from University of Roorkee, India; and a Ph.D. in civil engineering with specialization in water resources from Colorado State University. Prior to joining the South Florida Water Management District, he worked as an assistant professor in the Department of Civil Engineering at Colorado State University, where he taught courses in hydrology and water resources and conducted research in stochastic hydrology. In addition, he has taught courses in water resources at George Washington University, Washington, DC, and at Florida Atlantic University, Boca Raton, Florida. During his career, Dr. Obeysekera has published numerous research articles in refereed journals in the field of water resources. Dr. Obeysekera has over 20 years of experience practicing water resources engineering with an emphasis on both stochastic and deterministic modeling. He has taught short courses on modeling in the countries of the Dominican Republic, Colombia, Spain, Sri Lanka, and the U.S. He was a member of the Surface Runoff Committee of the American Geophysical Union and is currently serving as a member of a Federal Task Group on Hydrologic Modeling. During the last 17 years, Dr. Obeysekera has worked in south Florida as a lead member of a modeling team dealing with development and applications of computer simulation models for Kissimmee River restoration and the restoration of the Everglades Ecosystem. Currently, he is serving as the director of the Hydrologic and Environmental Systems Modeling Department at the South Florida Water Management District.

John Pitlick is an associate professor at the University of Colorado, Boulder, with research interests in surface water hydrology and fluvial geomorphology. He earned a Ph.D. from Colorado State University. His research examines processes of sediment transport and channel change in both natural and altered river systems. The principal goal of this research is to develop process-based models coupling hydrology, geomorphology, and material transport across a continuum of scales. Additional research being done in collaboration with fisheries biologists and aquatic ecologists seeks a more detailed understanding of interactions between geomorphology and ecosystem processes, including food-web dynamics and nutrient cycling. Most of this research is field based, with a geographic emphasis on rivers in the western United States.

Clair B. Stalnaker has been a key player in the instream flow arena for over 30 years—in research, method development and implementation, and policy. He organized and served as leader of the Cooperative Instream Flow Service Group (and various subsequent titles) under the U.S. Fish and Wildlife Service. This program brought together an interagency group of multidisciplinary scientists for the purpose of advancing state-of-the-art science and elevating the field of instream flow to national and international prominence. The primary focus of this group has been toward a holistic view of river science addressing the major components of instream flow management, namely hydrology, geomorphology, water quality, aquatic biology and connectivity and promoting instream flow regimes (incorporating intra- and inter-annual variability). He retired as a senior scientist with the U.S. Geological Survey where he was chief of the River Systems Management Section, Midcontinent Ecological Science Center, Fort Collins Colorado. He served as assistant professor of fisheries and wildlife science (1966 to 1976) and adjunct professor in the Department of Civil Engineering at Utah State University and has served as adjunct professor in the Departments of Earth Resources and Fisheries and Wildlife, Colorado State University. He has served on national and international technical committees and task forces and authored numerous publications focusing on the instream flow aspects of water allocation and river management. He served on the WSTB Committee on Western Water Management and is a member of the Science Advisory Board for the Trinity River Restoration Program, California.

Gregory V. Wilkerson is a visiting research assistant professor in the Department of Civil and Environmental Engineering at the University of Illinois at Urbana-Champaign. Dr. Wilkerson's research interests include research and development of solutions to water resource problems, multi-disciplinary approaches to stream restoration, river mechanics, sedimentation and erosion, environmental hydraulics, engineering hydrology, and statistics. His current research includes analytical model development for bioengineered systems, physical modeling of rivers, and quantifying the impact of increased water discharges in ephemeral channels. Dr. Wilkerson is currently a principal investigator with the National Center for Earth-Surface Dynamics, a National Science Foundation science and technology center. Dr. Wilkerson earned a B.S. degree in civil engineering from the Georgia Institute of Technology in 1989, an M.S. (1995) and a Ph.D. (1999) in civil engineering from Colorado State University.

Przemyslaw A. Zielinski is a senior quantitative analyst with Ontario Power Generation. As a senior analyst, he develops quantitative and qualitative methodology and analytic tools in assessing operational risk for Ontario Power Generation. Previously he was involved in OPG dam safety assessments as a dam safety hydrologist, and then as a senior scientist at Ontario Hydro. He was also assistant professor at Warsaw Technical University in Poland. His expertise is in the areas of risk analysis, assessment, and management; applied probability, statistics and stochastic processes; decision making under uncertainty; linear and nonlinear optimization; modeling of dynamical systems; and hydrology and water resources management. He chairs the Committee on Dam Safety of the International Commission on Large Dams. Dr. Zielinski earned his masters in mathematics from the University of Warsaw and masters in civil engineering and Ph.D. in stochastic hydrology from Warsaw Technical University in Poland.